George Collings

A Practical Treatise on Handrailing

Showing New and Simple Methods for Finding the Pitch of the Plank, Drawing the

Moulds, etc.

George Collings

A Practical Treatise on Handrailing
Showing New and Simple Methods for Finding the Pitch of the Plank, Drawing the Moulds, etc.

ISBN/EAN: 9783744679046

Printed in Europe, USA, Canada, Australia, Japan

Cover: Foto ©berggeist007 / pixelio.de

More available books at **www.hansebooks.com**

A PRACTICAL TREATISE

ON

HANDRAILING

SHOWING

NEW AND SIMPLE METHODS FOR FINDING THE PITCH
OF THE PLANK, DRAWING THE MOULDS, BEVELLING,
JOINTING-UP AND SQUARING THE WREATH

BY

GEORGE COLLINGS

AUTHOR OF "CIRCULAR WORK IN CARPENTRY AND JOINERY"

SECOND EDITION, REVISED AND ENLARGED

TO WHICH IS ADDED

A TREATISE ON

STAIRBUILDING

With Plates and Diagrams

Capio Lumen

LONDON

CROSBY LOCKWOOD AND SON

7, STATIONERS' HALL COURT, LUDGATE HILL

1890

LONDON:
PRINTED BY J. S. VIRTUE AND CO., LIMITED,
CITY ROAD.

PREFACE

TO THE FIRST EDITION.

THE Author of the following treatise having been employed for several years as a practical Handrailer, and having found great difficulty in applying the methods given in the various works which have been published on the subject, has endeavoured to work out a practical method of his own, which he now presents to his fellow-workmen, with a hope that it may meet with their acceptance, and be found of practical utility in the execution of this difficult branch of Joinery.

In compiling this work, the Author has consulted all the best authorities on the subject, and examined every source from which help was likely to be obtained. While therefore he does not lay claim to be the inventor of the whole of the system which is here given, for much of which he is indebted to those who have preceded him, he nevertheless believes that the simple method of finding the direction of the major axis for drawing the curves of the moulds has never been introduced before, and will be found to be much simpler as well as more practically useful than the equivalent method of finding the *oblique plane*, *pitching plane*, or *plane of the plank*, of the other systems. The method which he has introduced of drawing the curves is also considered to be much simpler than the old system of ordinates. The mould can be got out by the one method in every case without any variation.

Of course these eight Plates do not embrace all the different plans which may occur in practice, but they are representative of the majority of them; and those who master this system will have no difficulty in applying it to any other cases with different arrangement of plan.

It is believed that this treatise will be found sufficiently complete; the process of bevelling, jointing-up, and then squaring the *wreath* being fully explained, so as to leave nothing further to be desired except that all-important thing necessary to efficiency, namely, *practice*.

The Author has constantly applied this system himself in the practical execution of some of the best examples of Handrailing, and is therefore able to vouch for the correctness of the rules here laid down for getting out the moulds, by the results he has been able to obtain.

GEORGE COLLINGS.

PREFACE

TO THE SECOND EDITION.

———•———

A NEW edition of this little work being called for, some important additions have been made by the Author which will render the application of his method practically unlimited. Various new articles and diagrams have been introduced where such were considered desirable, including a new Plate (No. IX.) of a somewhat complicated character, with the necessary explanations and instructions. These additions, it is hoped, will considerably enhance the value and utility of this part of the present edition.

A treatise on STAIRBUILDING, which is the companion subject of HANDRAILING, has also been added to the present edition. On this subject the Author does not claim to have said much that is new. He hopes, however, that the matter in question will be found to be presented in an intelligible and acceptable manner, so as to be easily understood by those hitherto unacquainted with the subject.

LONDON, 1890.

CONTENTS.

PART I.—HANDRAILING.

PLATE I.

DRAWING ELLIPSE AND SCROLL.

PLATE II.

HANDRAIL OF STAIRS WITH STRAIGHT FLIGHTS.

PLATE III.

HANDRAIL TO QUARTER SPACE OF WINDERS.

PLATE IV.

WREATHS FOR A QUADRANT OF WINDERS.

PLATE V.

WREATHS FOR HALF-SPACE OF WINDERS.

PLATE VI.

WREATH FOR OBTUSE ANGLE WITH WINDERS.

PLATE VII.

MOULDS FOR SCROLL SHANKS.

PLATE VIII.

SCROLL SHANKS FOR WINDERS.

PLATE IX.

PART II.—STAIRBUILDING.

PRELIMINARY.

CHAPTER I.

PLANS OF STAIRCASES.

CHAPTER II.

THE WREATHED STRING.

CHAPTER III.

BRACKETTED STAIRS.

PART L

HANDRAILING.

HANDRAILING.

PLATE I.

DRAWING ELLIPSE AND SCROLL.

1. FIGURE 1 shows a practical method of describing a semi-ellipse on the principle of the trammel, but without making use of that instrument.

Draw the diameters, or *axes*, of the proposed ellipse, intersecting each other at right angles in the point I. Then take a slip of stiff paper, or a thin straight edge (as Fig. 1*a*), rather longer than the half major axis of the ellipse, and mark the point A near one end of the slip. Take thereon the length A B, equal to half the minor axis, and the length A c, to half the major axis of the ellipse. In order to find points through which to draw the curve, place the point B of the straight edge on the major axis (Fig. 1), and the point c on the minor axis. Then by carrying the slip round with the points B and c always on the axes, and marking a point at A each time the slip

is moved, any number of such points in the curve
may be found.

2. Suppose, however, that instead of the diameters, or axes, of the ellipse being given, we
have two points on the curve, as A and D, and also
the centre I and the length of the minor axis only
given, from which data it is required to determine
the length and direction of the major axis.   Then
the foregoing method of drawing an ellipse sug-
gests a ready way of doing this.   For instance,
provide a *square*, as H J K L (Fig. 1b), of any stiff
thin material, and at or near the middle of the
side J L, draw the line I G at right angles to J L.
Place this square with the point I to the given
centre I of the required ellipse (supposing of
course that the diameters and curve are not there),
and lay the line J L of the square as nearly as
possible in what is *supposed* to be the direction of
the required major axis.   Now apply A B, the
given half minor axis (on the slip Fig. 1a), from
the two given points A and D, making the points
B on the slip to lie upon the edge J L of the square
at B and E.   Prolong A B and D E from each point
to cut I G in C and F.   Then if A C and D F are
equal, J L is the direction of the major axis, and
A C or D F will be its half length.

If at the first trial it is found that A C and D F
are not equal, then move the edge J L of the

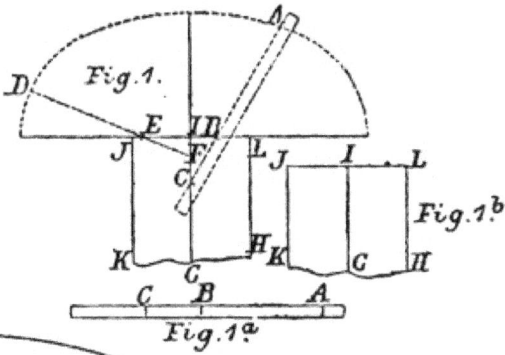

Fig. 1.

Fig. 1.ᵇ

Fig. 1.ᵃ

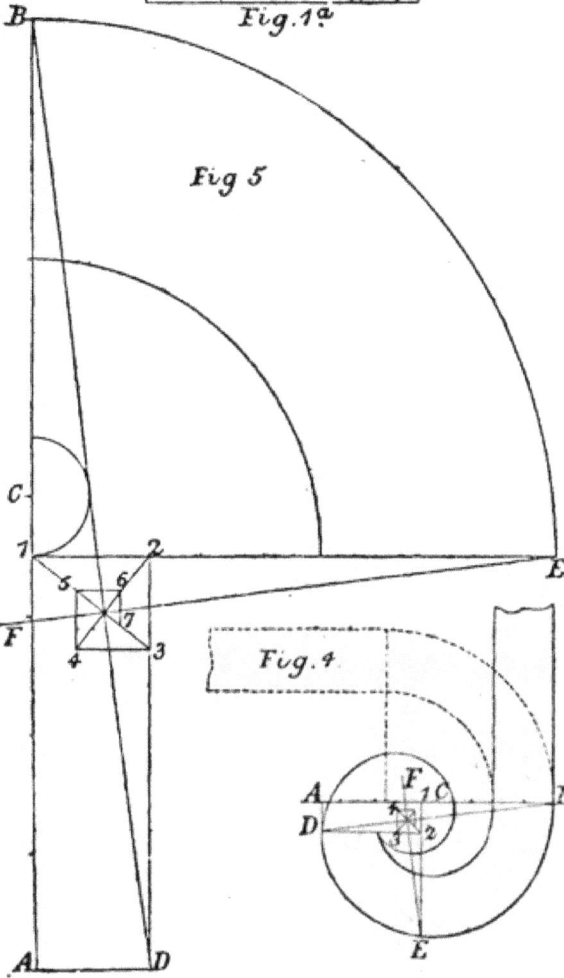

Fig 5

Fig. 4.

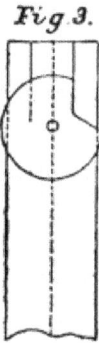

Fig. 3.

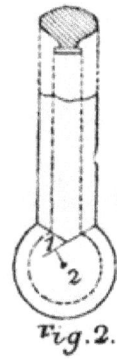

Fig. 2.

PLATE I.

square round until they become so, keeping I to
the given centre of the ellipse, B over J L, and A
to the given points.

An ordinary rule, such as is used by joiners,
will serve the purpose of finding the direction and
length of the axes.   And for a square, if for use
in a joiner's shop, nothing better can be had than
a wide piece of thin board with a line drawn in
the middle at right angles to one end.

We will suppose that Fig. 1 is drawn to a scale
of $\frac{3}{4}$-inch to a foot, when the major axis will measure
2 feet, and the minor axis 1 foot 4 inches, half of
which will be 12 inches and 8 inches respectively.
Lay down the centre and the two points A and D
at their relative distances from each other, full
size.   Place the square with the end of the line
in the middle to the given centre, then apply a
rule from the points A and D, being careful to let
each point be in an exact line with the edge f
the rule.   Lay the square so that 8 inches
measured from each point, as directed, shall be
exactly over the end J L, and the rule will cut
the line in the middle at 12 inches from A
or D.

If then a line be drawn across the end J L of
the square for the direction of the major axis,
and the curve described as before stated, it will
be found to pass through the two given points A

and D. Thus when these two lines A c and D F,
drawn as directed, are found equal in length,
each will equal half the longer diameter or
major axis, and the end of the square will give its
direction.

The method described above is the one we
propose to use for drawing the moulds throughout
this system of handrailing. It is very simple,
sufficiently correct, and can be applied with very
little trouble.

We shall explain hereafter how to find the
centre and the two points A and D, and we know
from the principles of geometry that the curve to
be drawn through the two points should be
elliptical, because they are in the circumference
of an oblique section of a circular cylinder.

The length of half the minor axis for the
elliptic curve will always be equal to the radius
of the quarter or half circle over which the two
points are situated.

3. Fig. 2 shows how to obtain the angle for
mitreing the rail into a newel cap. The section
on the edge of the cap should be the same as
the rail to be mitred into it. Draw a section
and plan of the rail, also draw the plan of
the circumference of the cap. Then take the
greatest distance to which the moulding is
worked on the under or upper side of the rail, as

the case may be, and draw a line parallel to the edge at this distance.

Draw also at the same distance another circle concentric with the outer circumference of the cap. Then from the intersection of the outside of the rail with the outside of the cap, and through the point where the inner parallel lines meet, draw the mitre. This will be found as near an approximation as possible without having recourse to a circular mitre to get both rail and cap to look alike.

If the mitre is carried in further than this, as is sometimes the case, the section of the cap will have to be found on this mitre line, and while the vertical projections of the moulding will remain the same, the horizontal will be increased considerably, producing a thinner appearance altogether on the cap than what the rail has.

4. Fig. 3 shows a method of cutting the mitre in the cap by using a piece of scantling one side of which is planed true. Draw a line in the centre of the width, and on each side of it set off the distance 1, 2 (Fig. 2). Then make a cut with a saw to these last two lines, and mark the width of the rail on the extreme edge of the cap, which should then be fastened to the scantling by driving a screw through the centre line into the centre of the cap.

Place one of the marks denoting the outside of the rail opposite one of the cuts, and make a saw cut to the required depth. Then turn the cap round until the other outside of the rail is brought opposite the other cut, and make another saw cut to complete the mitre.

Another method is to use one cut only in the piece, and to drive the screw through at the bottom of this cut into the cap at the point where the mitres meet. This plan is perhaps the least troublesome. The only objection to it is that the mitre cannot be cut quite home either way until the cap is released from the block, owing to the screw being in the way of the saw.

5. Figs. 4 and 5 show a handy method of drawing a curtail scroll. In Fig. 5, which is a magnified representation of Fig. 4, the width A B is divided into eight equal parts. Make A D equal to one of these parts and at right angles to A B, and join B D. Place one foot of the compasses at C, the centre of A B, extending the other foot to touch the line B D, and draw the arc, cutting A B in the point 1. Then 1 is the centre of the first or largest quadrant of the scroll. Draw 1 E square to A B, and from 1 as a centre draw the large quadrant B E. Draw D 2 parallel to A B. The point 2 is then the second centre. In order to find the remaining centres, from E draw E F square

to B D. From 1, through the point where B D
and E F cross each other, draw 1 3. From 3,
draw 3 4, parallel to 1 E. Then draw 2 4, and
from where 3 4 cuts this last, draw 4 5 parallel
to A B, 5 6 parallel to 1 E, 6 7 parallel to 4 5,
and so on until a sufficient number of centres are
found from which to complete the scroll.

The advantage of this method of drawing a
scroll is that the width A B may be varied to suit
any width of rail. For a wide one it may be
increased, and for a narrow one diminished. If
another quadrant or part of a quadrant is required,
it is only necessary to increase its radius in the
same proportion as the first radius is to the second,
and then set the foot of the compasses at this
distance from B, on the line A B, as shown by the
dotted lines on Fig. 4.

A scroll being once drawn as directed, the
width, A B, may be reduced or enlarged as required by using the same mould. Thus, mark off
the width by drawing a straight line of the required length across the scroll wherever it will
apply, and from the point where this line cuts
the outer curve draw a line representing the side
of the straight rail as a tangent to the curve, that
is, at right angles to the radius at that point.
In the scroll before us this point may be situated
anywhere (according as the length of line or

width determines) on the outer curve between B and D, and also beyond B on the dotted curve (Fig. 4).

Several other methods of drawing scrolls or spirals will be found in a treatise on " Practical Geometry for the Architect, Engineer, Surveyor, and Mechanic," by E. Wyndham Tarn, Architect.*

* Crosby Lockwood and Son.

PLATE II.

HANDRAIL OF STAIRS WITH STRAIGHT FLIGHTS.

———

6. This Plate presents a very simple case in continued handrailing. The lines necessary for this pair of wreaths might be obtained from the pitchboard. We think, however, that it will be better to leave the pitchboard alone for the present, and explain the method taught in the following pages as fully as possible in its application to this simple case, instead of leaving it until we come to a more complicated one.

Fig. 1 is the plan of a well-hole of a staircase, with straight flights above and below the landing. The rail is shown the whole width up to the springing line of the curved part; beyond that the centre only is laid down with the tangents to the half circle, 1, 2, 3, 4, 5, and the diagonal, 1 3. The riser lines, landing, and starting are not drawn to the centre of the circle: the distance from riser landing to 2, and from 4 to riser starting, is made equal to half the *going* or tread of one step. By this arrangement the rail has a better appearance than when more of it is thrown on the level.

Fig. 2 is the development, or stretch-out in elevation, of the centre line and tangents of Fig. 1. Draw the landing with a step above and below, erect the perpendiculars to coincide with the points 1, 2, 3, 4, 5, of Fig 1. The distances from 1 to 2, from 2 to 3, and so on, will then be equal in each figure. Draw the under side of the rail resting on the angles of the flyers, and at the distance of half the thickness of the rail draw the centre line, and continue the upper one downwards to cut the vertical line 4 at F; continue the lower one upwards to cut the vertical line 2 at B, and draw the horizontal line B F. The centre joint at c will be vertical, owing to B F being level. Draw the upper and lower joints square to the centre line. From A, where the centre pitch cuts the vertical line 1, draw A E horizontal. From E set off E D equal to the diagonal 1 3 (Fig. 1). Draw D c, then this line will be the diagonal for the mould, and when placed to its proper pitch on the plan, will lie over 1 3 from end to end.

Fig. 3 shows the method of drawing the mould. Make A B (Fig. 3) equal to A B (Fig. 2), and draw B c at right angles to B A and equal to B c (Fig. 2). Then A c, the hypothenuse of the right-angled

PLATE II.

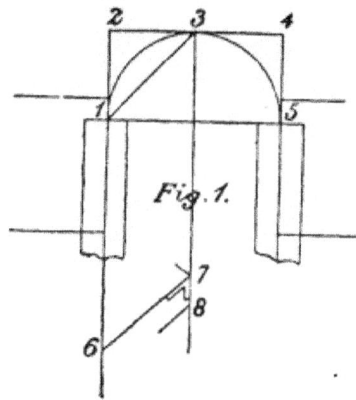

Fig. 1.

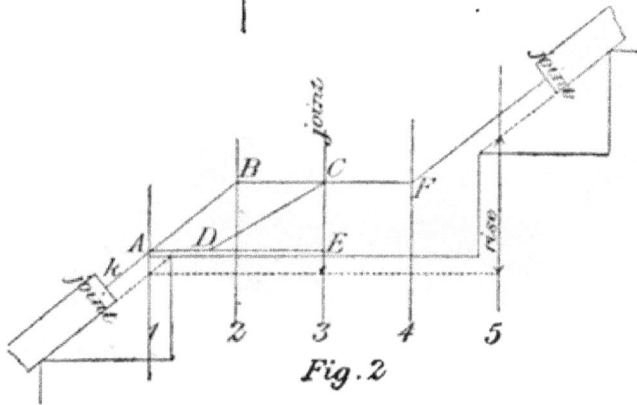

Fig. 2.

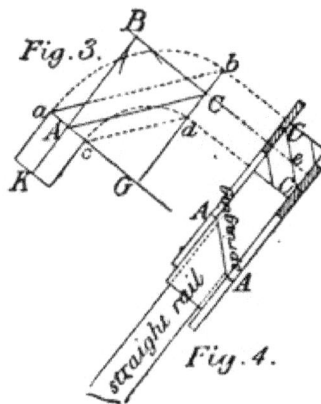

Fig. 3.

Fig. 4.

triangle A B C, is equal to D C (Fig. 2). Prolong the
line A B until the distance A K (Fig. 3) is equal
to the distance A k (Fig. 2). Draw the end of
the mould square to A B; through A and C draw
A G and C G parallel to B C and A B respectively; G
will then be the centre of the elliptic curves of
the mould. The line B C being level, the
angle B will be square, the bottom end of the
mould will therefore be square, and the same
width as the rail. Draw the width of the rail
parallel to A B. To find the bevel and width for
the top end, take A B (Fig. 3) in the compasses,
and with one foot at any point 6 (Fig. 1) on
centre line of rail, draw the arc cutting the centre
line of well-hole at 7. Draw 6 7, and this
gives the bevel. Parallel to 6 7 set off half the
width of the rail, and 7 8 will be half the width
of the mould. Repeat this distance on each side
of C (Fig. 3), and this gives the point to which we
have to draw the quarter-ellipses of the mould.
Or, $a c$ being the width of the rail, the lines $a b$,
$c d$, drawn parallel to A C, will give $b d$ its width at
the centre of the wreath; and G a, G b will be the
semi-axes of the ellipse for the outside of the rail,
G c and G d the semi-axes of the ellipse for the
inside. B C being level, A G and C G will be the
directions of the diameters. G is the centre
coinciding with the centre of the half circle in

Fig. 1. Therefore from G to the inside and outside of the mould on each diameter will be half the shorter and half the longer axes respectively. If now we mark off these distances from the end of a thin straight edge, and carry it round, keeping one of the points on each diameter (as shown by Fig. 1, Plate I.), any number of points can be obtained through which the curves of the mould may be correctly drawn. Draw the lines A B and B C opposite each other on both sides of the mould, and square the springing A G across the inside edge. This will mark the termination of the straight part and the commencement of the circular, and will be made use of when the mould is applied to the plank. Both the half wreaths being alike, there is no necessity to get out a separate mould for the upper one.

Fig. 4 shows the application of this mould to the plank for the purpose of working it to the twist or *bevelling* (we shall throughout this work call it *bevelling*). The piece is first cut out square to the shape of the mould, leaving the stuff full all round, and one side planed true. It will be found better to have two moulds for bevelling, one for each side of the piece, to supply the place of the line that will be lacking by reason of the piece being cut out square.

In Fig. 4 there are two moulds shown with the

piece between them, as it would be in practice for bevelling. It is supposed to be placed with its convex or outside edge downwards, and we are looking at the concave or inside edge, the end or centre joint c and the thin concave edges of the moulds, the upper ends of which are shaded.

The lines A B and B C should be drawn on both sides of the piece, to coincide with those on the mould when the edges of both are even. Square c across the centre joint, as shown by the dotted line, and through the centre e of the thickness draw the bevel c c, the same as found at 6 7 (Fig. 1). From c c (bevelled line) draw another line on the face of the piece parallel to B C. Slide the mould on the top side upwards, so that the point c may be moved from the square dotted line until it lays over c c, the bevelled line, keeping A B on the mould over its corresponding line on the piece.

Slide the mould on the under side in a similar manner the reverse way, so that the point c may be under the bevelled line c c. Fasten both moulds to the piece with small screws, so that the holes made by the screws may be taken out in squaring the wreath. Take off the superfluous stuff inside until a straight edge applied in a perpendicular direction will touch the edges of the moulds all round. Take off the stuff outside similarly, leaving it a little full.

Before taking the moulds off the piece, draw the line marked "springing" from the point where A G is squared across the edge of the mould. This line is made use of when jointing the wreath to the straight rail, and should always be perpendicular when the wreath is placed to its proper inclination.

Both wreaths being bevelled as above, the easing may be roughly made, that is, some of the superfluous stuff may be taken off at top and bottom; or, in other words, the wreaths may be roughly squared, but should not be finished until the joints have all been made and bolted together. For the lower joint a bevel should be set with its stock to the under side of the straight rail (Fig. 2), and the blade made to coincide with the vertical or springing line, 1 A, or the hypothenuse and rise of the pitchboard. Then, in applying this bevel, hold the stock to the under side of the straight rail, and make the joint so that the springing line on the wreath shall coincide with the edge of the blade.

The top joint is made in a similar way, and by using the bevel as directed, the springing line will be brought to its proper direction when the joint is bolted together. Or the pitchboard may be applied with the rise to the springing, and the line of the straight rail marked on the side of the wreath by the hypothenuse.

This way of obtaining the bevel will be found to apply in every case with or without winders, the stock being held to the under side of the straight rail and the blade made to coincide with the springing line on wreath.

To make the centre joint, a three-inch plank should be used, one side planed true and the edges shot square. The width should be equal to the distance between the inside edges of the rail in Fig. 1. Ascertain how much the rail rises on the under side from 1 to 5 (Fig. 1), as shown on Fig. 2. From the point where the under side of the rail cuts the vertical line 1 A, square over to the vertical line 5 ; then from this point to that where the under side of the upper rail cuts the same line will be the rise of the rail on the springing from 1 to 5 (Fig. 1). Mark this rise on the face of the plank. Take the first wreath, and plane the joint square to the line B C and face of the stuff. Handscrew this wreath to one edge of the plank, making the springing line to coincide with the face. Place the point where the under side of the lower rail cuts the springing to the bottom line on the plank. Then take the top wreath and hold it to the opposite edge of the plank. Keep the springing to the face, and fit the joint until the point where the upper rail cuts the springing is brought to the second line, or

the rise on the plank. This process may appear
tedious, but in practice it gives very little

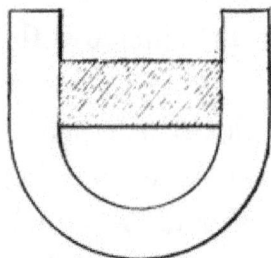

Fig. A.

trouble, while it insures
perfect accuracy in the
result.

The annexed figures, A
and B, show the plank with
a pair of wreaths as applied
to it when jointed up at
the centre joint. Fig. A
is the plan with end of plank shaded and the
wreaths going round the face in a semicircle.
Fig. B is an elevation of the face side of the

plank with the wreaths in
position, the straight part
of the lower wreath on its
under side being placed to
the line a, and the straight
part of the upper wreath—
also on its under side—
being placed to the line b,
the distance between a and
b being equal to the rise as
shown in Fig. 2, Plate II.
This will be found to be
a most expeditious and

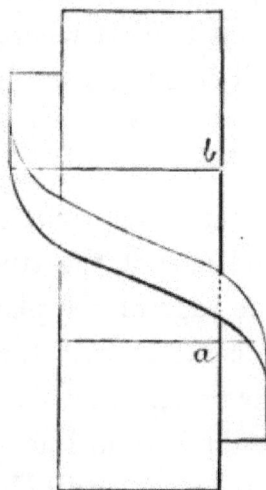

Fig. B.

correct method of making the centre joint in a
pair of wreaths, however large or small the well-

hole may be. It is the only joint presenting any difficulty in a plan of this kind.   The straight-ends, it will be seen, are made to lie in the right direction, both sideways, and on the top and under sides. Besides which, if the end for the joint of the lower wreath is planed true, and the piece be then hand-screwed to the edge of the plank as directed, the end of the upper one can be fitted to it with an exactitude not to be surpassed by any other method.

For large circular well-holes where the wreath may be in several lengths it is also the safest and best plan to make (roughly but truly) a cylinder, or semi-cylinder, as the case may require, and of sufficient length.   The heights should then be set out on this in the manner similar to that given in the foregoing directions.   By this means the ends of wreaths beyond the easings at floors and land-ings may be made to lie in a true and correct position, and to their proper heights.

This method of determining the rise of the rail on the springing line will be found to apply in every case.   In making the top and the bottom joints the centre of the rail is supposed to be placed to the centre of the wreath ; this, however, may be varied either way a little, so as to bring the rise or height correct.

To obtain the length of the rail, a rod about

2 inches wide should be used; this should be laid
edgewise on the nosing line of the flyers, or
parallel to them, and the springing or vertical
line marked on top and bottom. Where a scroll
is used, the face of second riser at bottom and
springing at top should be marked on the rod.
If we have a scroll and two or three commode
steps, then the face of third or fourth riser, as the
case may be, and springing. Then this rod
should be held to the under side of the rail when
jointing up, and the springing lines on the wreath
and those on the rod made to range in the same
perpendicular direction.

It may perhaps be thought that we have
entered somewhat too minutely into details for
such a simple case as the above; but we have
done so because the method of proceeding as here
explained will be found to apply throughout in
every case, except that a slight variation will be
necessary for finding the centres in obtuse and
acute angles; and when it is desirable to take in
more than a quarter of a circle.

7. In the example we have just been considering
(Plate II.) the rise and going of the treads and
the width of the well-hole are such that the
raking centre lines of the rail cut the perpendiculars 2 and 4, Fig. 2, in points B and F at the
same distance above the landing, and by joining

these points as directed, a perfectly horizontal line is produced which is the one required. This may not be the case once in ten, or even a larger number of times, with plans of a similar description. If the riser lines of this plan were placed at the springing, the point B in Fig. 2 would be higher and point F lower than is there shown to be the case; and if these two points were so placed and connected by a straight line, it would be considerably out of the horizontal, and consequently the pieces for the wreath would be thrown out of their proper position. A novice in this kind of work finding this to be the case might be at a loss to know how to proceed, and might probably condemn the system as being incorrect.

There is, however, a method—illustrated in the accompanying Figs. C, D, E, F—by which the correct position of the risers may be ascertained to a very great nicety. But let it first be observed that the raking centre lines of the straight rail should always cut the central horizontal line, drawn at its proper height above the landing, in the perpendiculars 2 B and 4 F, as shown in Fig. 2, Plate II.

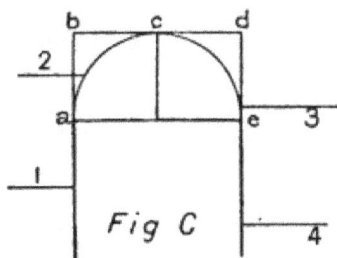

Fig C

In Fig. c let the semicircle *a c e* be the plan of
the centre of the rail, and enclosed by the tangents
a b, b c d, d e.   Unfold the tangents as at *a b c d e*
on Fig. D.   Draw the landing line, *f g,* and at
the proper distance above it (this distance should

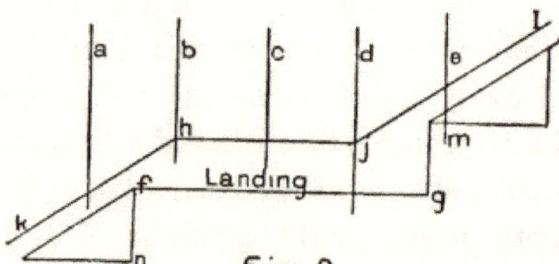

Fig D.

be about three or four inches to the under side of
the rail), the centre line of the wreath, *h j,* cutting the perpendiculars *b* and *d* in points *h* and *j.*
From *h* draw *h k,* and from *j* draw *j l* to the rake
as given by the hypothenuse of the pitch-board.

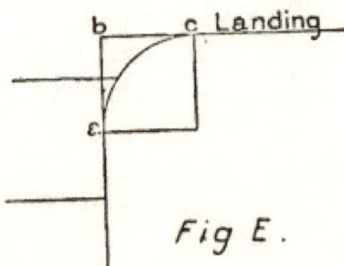

Fig E.

From *h k* and *j l* set
off half the thickness
of the rail downwards,
and draw the parallel
lines as shown.

Then apply the rise
of the pitch - board
from the landing, as at *g m,* to cut the under
side of the upper rail in *m;* and *g m* will be
the proper position for riser 3 on plan Fig. c.
From *f* let fall the perpendicular *f n,* and this

line will give the correct position of the riser 2
on Fig. c.

This method will give the position of risers in
all plans of a similar description. In some there
will be found no need to use it. In others it will
not only give the position of the risers as indicated, but it will also improve the appearance of
the rail. No more trouble need be experienced
in making the stairs than if the risers were placed

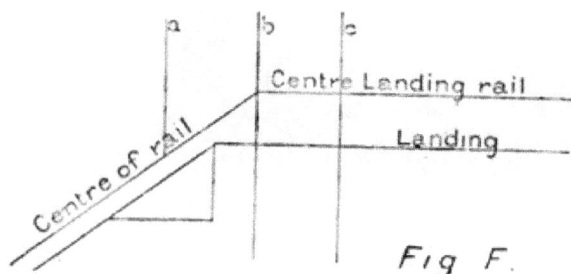

Fig F.

at the springing lines, and an altogether better
looking job may be the result.

Figs. E and F show the same method as applied
to finding the position of the last riser at landing
on to a floor. The similarity of those two Figs.
to the first or lower half of Figs. c and D is so
manifest that a detailed description is not considered necessary.

In starting from the level, as off a floor on to a
flier, the upper half of Figs. c and D would apply.
The procedure in both cases being exactly the

same, separate figures and description need not
be given.

The thickness of stuff necessary for wreaths
may be stated generally to be the width of the
rail. Thus, for a rail 3 in. by $2\frac{1}{2}$ in. the plank
for the wreaths would require to be 3 in. thick.
Where there is an easing on the upper or lower
end of the rail, as in Plates III., IV., V., and VI.,
stuff of the same thickness as the rail is wide will
also be required for cutting them out of. This
will be found to be the best rule to follow
generally, except for a quantity of straight rail,
as nothing need be wasted by cutting even a perfectly straight piece of rail out of the extra thickness of plank. It is sometimes said that the
plank for the wreaths need be no thicker than
the rail. This, perhaps, may be near enough in
work of very little importance, and where the
workman may be contented to give the wreaths
a crippled and unsightly appearance, as he is
sometimes compelled to do for the sake of saving
a very small quantity of material. This, however, should never be done except where the
work is very indifferent indeed, or stuff of the
requisite thickness cannot be obtained readily
when wanted. But when this can be procured
it should be at the least half an inch thicker than
the rail. By this means there will be a little

stuff to spare, whereby the easing in the rail may be continued along the wreath as far as may be necessary to make the whole assume a graceful form and look pleasing to the eye. Cases sometimes occur where it is necessary to have the plank as much as an inch thicker than the rail. In some instances the wreath has a very steep pitch running up in an almost perpendicular direction over the ends of winders that have a very small well-hole.

For cases of this description, and also for scroll eyes and shanks where there are commode steps, extra thickness of stuff will be absolutely necessary. It will also be the cheapest in the end, as by using it a far better looking result may be produced, and much time saved in the working.

The matter, however, is one that must be left to the judgment of the individual workman to a great extent, as no hard and fast rule can be laid down. But the foregoing remarks will be found to apply generally as a guide.

CLASSIFICATION OF MOULDS.

8 Perhaps it may help to render the study of the following Plates more interesting, and make them more easy of comprehension by the uninitiated, if we state that every mould according to this system will be found to belong to one of three

classes, with an occasional exception similar to
that shown in Plate IX.

The first and simplest is that represented on
Plate II. All that is necessary to produce a
mould of this class is to lay down the centre lines
A B and B C at right angles to each other, and
parallel to them C G and A G (Fig. 3). If then
we mark off half the width on each side of A and
C as directed, A G and C G at once give the direction
and lengths of semi-diameters for drawing the
curves of the mould.

The accompanying Fig. P is a facsimile of
Fig. 3, Plate II., without the
diagonal and curves. A B and
B C are the centre lines, and A G
and C G the diameters. The top
mould on Plate V. and the mould
for scroll shank on Plate VII.
also belong to this class.

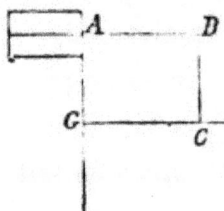

Fig. P.

An example of the second class, and almost as
simple as the foregoing, is supplied by the bottom
mould of Plate IV. This kind of wreath will
never present any difficulty. The diagonal A C is
drawn first (Fig. 3). Then A B and B C, and
parallel to these A K and C K. Then the direction
for the longer diameter will always be parallel to
the diagonal A C through the centre K. Having
obtained these lines, we next find the half widths,

and lay them down on each side of A B and B C. This gives the points M, N, O, P, to which the curves of the mould should be drawn.

The annexed Fig. Q is a facsimile of Fig. 3, Plate IV., without the curves, A B and B C being the centre lines, C K and A K are parallel to them. K is the centre of the elliptic curve.

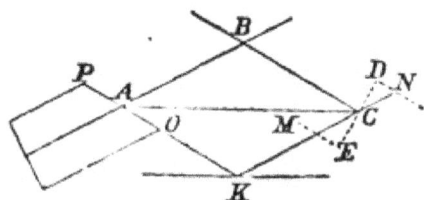

Fig. Q.

The line parallel to A C through K is the long diameter. M, N, O, P are the points on C K and A K, to which the ellipses should be drawn, and D E is the joint line. Figs. 3 on Plates IV. and V. belong to this class.

The third class is represented on Plate III., and this is the only kind of mould where the use of the square is necessary for finding the direction of the diameters. The same method is followed as in the last example until the centre of the curve is obtained. We then apply the square, and find the direction as explained.

The accompanying Fig. R is a facsimile of Fig. 3, Plate III., without the curves, &c. A C is the diagonal, A B and B C the centre lines; N W and S W are drawn parallel to them. M, N, P, S are the points to which the curves should be drawn. w is the centre of the ellipses, and D E the direction of the longer diameter. Fig. 4, Plate IV., Fig. 3, Plate VI., and Fig. 3, Plate VIII., belong to this class.

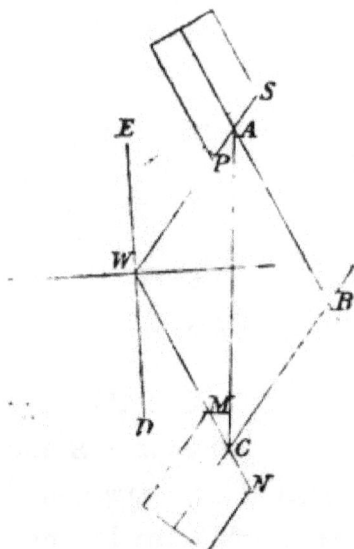

It will be seen that the finding of the direction of this major axis, or longer diameter of the ellipse, is the same thing as finding the *oblique plane, pitching plane,* or *plane of the plank* in other systems of handrailing, and is done in a more easy and less intricate manner. In other systems the plan and mould figures have to be connected in a particular way, so as to obtain the required pitch for the plank. In this system, on the contrary, each figure is separate and distinct, thus rendering them clearer and more easily understood. And if we have succeeded in making the use of the square for determining the direction of the diameters understood, no difficulty whatever should be experienced in describing the moulds.

If we examine Fig. P somewhat further, and

Fig. R.

compare it with Figs. Q and R, we shall see that
this simple rectineal figure contains every essen-
tial requisite for the moulds as here constructed,
and also that the same general method is ad-
hered to in the projection of all three, and therefore throughout this entire system of lines.

Thus A B and B C are similar lines in all three
figures, and lie over exactly the length of their
respective plans from end to end. So also of the
parallels to those lines. If a diagonal were drawn
from A to C in Fig. P, it would be similar to and
answer the same purpose as the diagonals of the
other two figures. The bevels and widths are
obtained similarly in all three figures (those for
Fig. P being exactly the same in principle as for
the other two). The centres G, K, W are all perpendicularly over the centres of their respective
plans. And lastly, the curved edges of the moulds
are all drawn to and finish against the parallels
to A B and B C.

PLATE III.

HANDRAIL TO QUARTER-SPACE OF WINDERS.

———

9. This plate shows the method of getting out the mould for a handrail where there is a quarter-space of winders connecting straight flights.

Fig. 1 represents the plan of such an arrangement, and the whole width of the rail is laid down to the springing lines of the curve. In the quarter circle the centre only of the rail is drawn, and this is enclosed by the tangents 1 2 and 2 3.

Fig. 2 is the development, or stretch-out in elevation, of the centre line and the tangents of Fig. 1. Draw the vertical lines at the same distance from each other as the points 1, 2, 3 of Fig. 1; also the risers as they occur on the centre line, and the tangents of the plan, with the flyers above and below.

The pitches for the straight rail are drawn as previously described, the under side resting on the angles of the flyers, and the centre at the distance of half the thickness of the rail. Continue the centre of the upper rail in a straight line to

PLATE III.

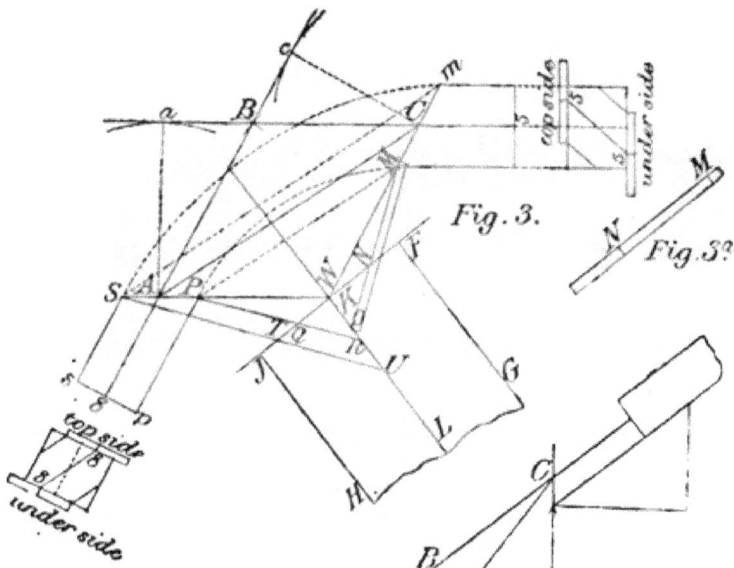

Fig. 3.

Fig. 3.?

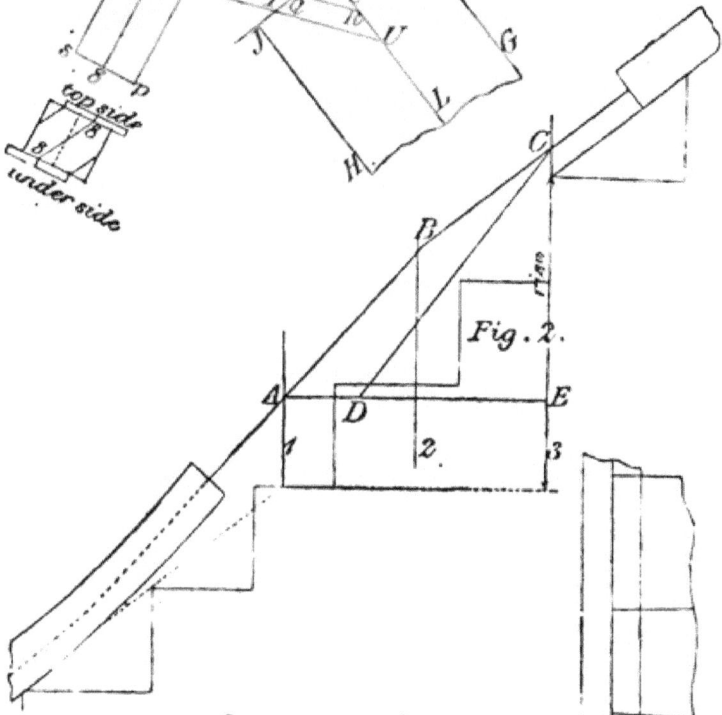

Fig. 2.

Fig. 1.

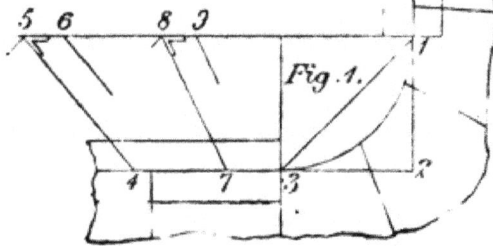

o 3

B, and from B continue it downwards to meet the centre of the lower rail, so as to form a fair easing on the same. In cutting out this easing, the stuff should be left full all round until the wreath is jointed to it.

The under side of the lower rail is prolonged to the vertical line 1 A (Fig. 2), as shown by the dotted line. From the point where this dotted line cuts 1 A square over to the springing, 3 c. Then from the point where this square line cuts 3 c to that at which the under side of the upper rail cuts the same line, will be the rise or height at which the wreath should be jointed to the lower rail.

The diagonal for the mould is obtained as before described. Thus from the point A, where the centre pitch cuts the springing, draw A E square to 1 A. From E set off E D equal to 1 3 (Fig. 1), and draw D C. Then D C will be the diagonal sought.

In order to get out the mould, we proceed as described before. Make A C (Fig. 3) equal to D C (Fig. 2). Take A B (Fig. 2) in the compasses as a radius, and with one foot at A (Fig. 3) draw an arc. Again, take as a radius B C (Fig. 2), and with one foot at C (Fig. 3), draw another arc cutting the former one at B. Draw A B and B C, and prolong them so as to make the distances A 8 and C 5, from A and C to the ends of the mould, equal to the distances from A and C to the joint in Fig. 2. Draw the end of the mould square to these lines. Through A and c draw A w and c w parallel to B C and A B. Then w will be the centre of the elliptic curves of the mould.

To find the bevel and width for the upper end of the mould, draw A *a* at right angles to B C, and with the distance A *a* as a radius, and the point 4 (Fig. 1) anywhere on the middle line of the rail as a centre, draw an arc cutting 1 5 at 5. Join 4, 5; then this will be the required bevel. Parallel to 4 5, set off half the width of the rail, and 5 6 will be half the width of the mould. Repeat this distance on each side of B C (Fig. 3), and draw parallel lines to cut the springing c w in the points M, *m*.

For finding the width of the lower end of the rail at A, draw c *c* at right angles to c w; and with the distance c *c* as a radius, and any point 7 (Fig. 1) on the centre of the rail as a centre, draw an arc cutting 1 8 at the point 8. Join 7, 8; then this will be the required bevel. Parallel to 7 8, set off half the width of the rail, and 8 9 will be half the width of the mould. Repeat this distance on each side of A B (Fig. 3), and draw parallel lines to cut the springing A w in s and P. Otherwise, by drawing *m*s and M P parallel to A o, we find the points s and P on the line A W. Then if we draw s s and P p parallel to B A, we get the width of the rail.

This gives the points to which we have to draw the curves of the mould, two for the inner and two for the outer ellipse. The centre w of these ellipses has already been found, and the length of half the shorter diameters or minor axes will be equal to the radii on plan (Fig. 1).

From these we have now to determine the direction and length of the longer diameters or major axes.

Let F G H J be the square for this purpose (as described in Plate I.), with K L at right angles to J F. Place this square so that the point K at the end of the line K L shall fall on the point w, the centre of the ellipse, and that J F shall lie in the *supposed* direction of its longer diameter. M and P are the two points on the inner curve of the rail through which the ellipse has to be drawn. Now let M N and P Q be half the shorter diameter (equal to radius for inside of rail on plan), and prolong M N and P Q to cut K L in the points o and R. Then if M o and P R are equal, J F will be the direction of the greater diameter, and M o or P R will be half the length of it. If at the first trial M o and P R are not equal, we must move J F round until they become so, always keeping the point K

to the centre w, and the points N and Q on the
edge J F of the square.

A thin straight edge, with the half shorter
diameter marked on one edge as shown at M N
(Fig. 3a), affords the most ready way of doing this.
Then M is placed alternately to the points M and
P, and N is made to lie on J F, which is placed so
that the continuation of the lines M N and P Q to
K L shall be equal. J F then gives the direction
for drawing the longer diameter. Or better still,
let an ordinary two-foot rule be held with the
edge to the points M and P, and the square so
placed that the distances of the point where the
rule cuts K L from M and P shall be equal, keeping
the points N and Q denoting the half shorter
diameter over J F.

To find half the longer diameter for the outside
curve, take the radius for the outside of the rail
on plan (Fig. 1) in the compasses, and placing
one foot at s (Fig. 3), draw an arc cutting the
longer axis at T. Draw S T U, then S U will be
the half longer axis sought.

It should be observed that the direction of the
diameter once found answers for the inside, the
centre, and the outside of the mould. It may
be found from any one of these three by em-
ploying the radius of plan corresponding to
whichever we may prefer using, which radius

will always be half the shorter axis for the elliptic curve.

Having found the direction of the axes, the process necessary to complete the mould is so simple as to need no further description.

The application of this mould to the plank for bevelling the wreath will be somewhat different to that on the preceding Plate, owing to there being two pitches and two bevels.

The piece is first cut out square to the shape of the mould, and the upper side planed true. The lines A B and B C are then drawn on the piece to coincide with those on the mould when the edges of both are even. Square these lines across each end, and through the centre of the thickness draw the bevels 5 5 and 8 8, the same as found at 5 and 8 (Fig. 1). Then from the points where the bevels across the ends cut the face, draw on the piece lines parallel to A B and B C.

The piece is now ready for bevelling, and for this purpose it is necessary to have two moulds. Place the mould on the upper side in such a position that A B and B C of the mould may be over, and in the same direction as the lines on the piece that are drawn parallel to A B and B C from the bevels across the ends.

The mould for the under side is applied in an exactly similar way, the lines of the mould being placed over those on the piece that are drawn parallel to A B and B C, the difference being that one is moved upwards and the other downwards.

If this is clearly understood, there should be no difficulty in bevelling this or any other wreath. The ends of the piece are shown at Fig 3, with the square and bevelled lines drawn on. The ends of the moulds are also shown in the position they should occupy for bevelling the wreath.

Fasten the moulds to the piece and take off the superfluous stuff until a straight edge applied in a perpendicular direction will touch the edges of the moulds all round, leaving the stuff a little full on the outside. Before taking the moulds off, mark the springing across the inside edge, as on the preceding Plate.

In jointing this wreath to the lower rail with the easing, the centres of each are supposed to be placed opposite to each other. This, however, may be varied a little either way, so as to bring the height correct.

The bevel is obtained as before described (page 14), the stock being held to the under side of the straight rail and the blade made to coincide with the vertical or springing line 1 A (Fig. 2). Then in making the joint, hold the stock to the under side of the straight rail, and fit the joint until the springing line on the wreath coincides with the edge of the blade. The top joint is made in a similar manner and with the same bevel.

The easings here, as in every other case, on both wreath and rail may be roughly made, but should not be finished until the joints have been completed and the parts bolted together.

PLATE IV.

10. This plate shows the construction of wreaths for a quadrant of winders, a quarter-space or landing, and straight flights above and below.

Fig. 1 shows the plan of the well-hole, enclosing the half circle with the tangents 1, 2, 3, 4, 5. Fig. 2 is the development in elevation of Fig. 1, the distances from 1 to 2, from 2 to 3, and so on, being the same in both figures. The riser lines are set out in the development as they occur on the centre line and tangents of Fig. 1.

Draw the under side of the straight rail resting on the angles of the flyers, and at the distance of half the thickness of the rail set out the centre line. Continue the upper one in a straight direction to cut the vertical line 4 at D, and from this point draw the centre downwards to meet the lower centre, so as to form a fair easing on the straight rail. Draw the joints square to the pitches. Continue the under side of the lower straight rail to cut the vertical line 1, and from this point

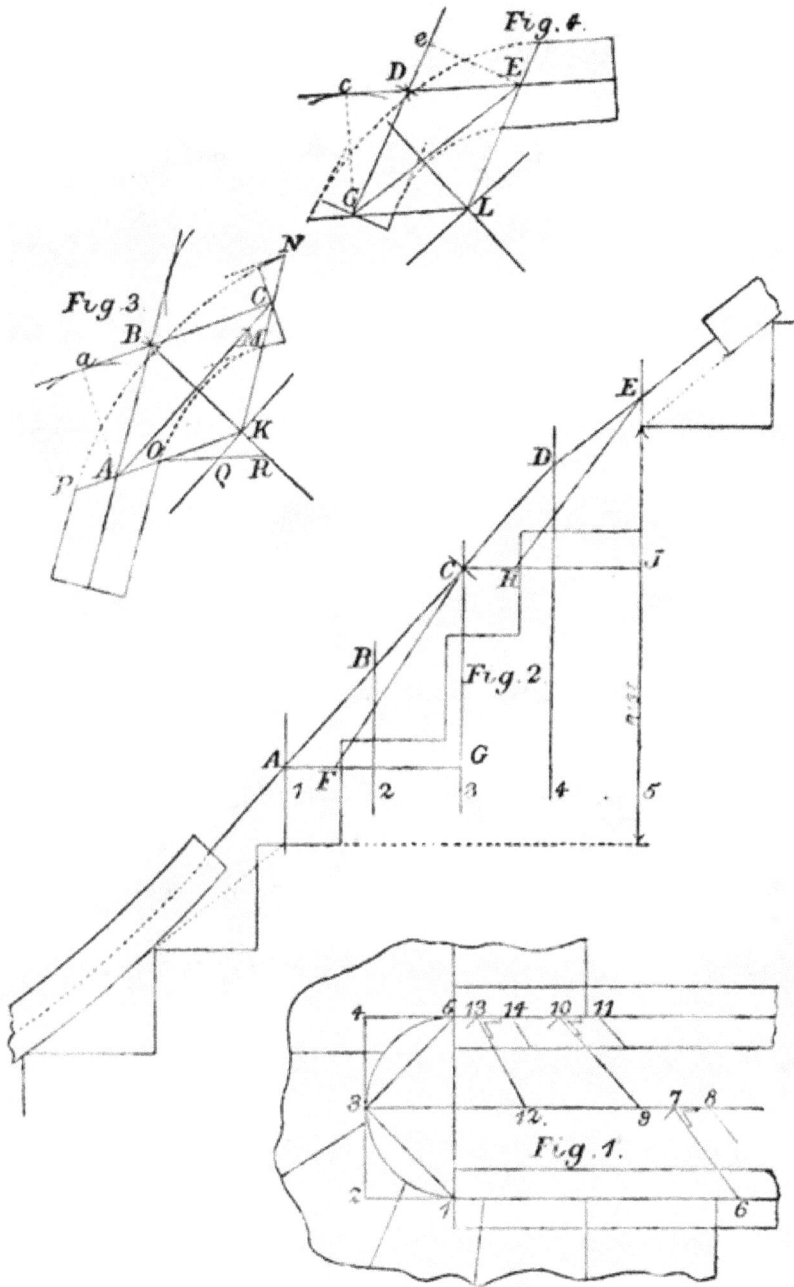

PLATE IV.

Fig. 4.

Fig. 3.

Fig. 2.

Fig. 1.

square over to the springing 5, thus obtaining the rise or height to which the wreaths should be jointed to the lower rail.

From A and c draw the horizontal lines A G and c J. Set off from G and J, G F and J H, equal to 1 3 or 3 5 (Fig. 1). Draw F C and H E (Fig. 2); these will be the diagonals for the moulds.

Fig. 3 shows the method of drawing the lower mould. Make A c equal to F C (Fig. 2); then take A B or B c (Fig. 2) in the compasses, and with one foot at A and at c (Fig. 3), draw arcs cutting each other at B. Draw A B and B c, and parallel to them draw c K and A K. Then K is the centre of the ellipse. Prolong A B, and make the distance from A to the end of the mould equal to the distance from A to the joint in Fig 2. Draw the end of the mould square to A B.

Both pitches of this wreath being alike, only one bevel is required. Draw A a at right angles to B c, and with A a as a radius and the point 6 (Fig. 1) as centre, draw an arc cutting the centre line at 7. Draw 6 7, and this gives the bevel. Parallel to 6 7 set off half the width of the rail, and 7 8 will be half the width of the mould. Repeat the distance 7 8 on each side of A B (Fig. 3), and draw parallel lines to cut the springing line A K in the points o and p. Also repeat the same distance on each side of B c, and draw parallels to cut C K in M and N. Then where these parallel distances to B C cut C K, will be the points at this end of the mould to which the elliptic curve should be drawn. M and N being these points, the outside elliptic curve will extend to N, while the inside curve will not reach beyond M. Thus taking the length of the mould from o to M and from P to N, it will be a quarter of an ellipse over a quarter of a circle inside and out. If this mould (leaving the curves for the moment out of consideration) were placed to its pitch over its plan (Fig. 1), the four lines forming the rhombus would lie over the four lines which on the plan form the square. A B would be over 1 2; B C over 2 3; A K and C K over the two opposite sides; A C over the diagonal 1 3; and if we were to draw the quarter ellipse in the centre of the mould from A to C, this quarter ellipse would lie over the quarter circle 1 3. Similarly the inside and outside quarter ellipses extending from o to M and from P to N would lie over the inside and outside quarter circles of the rail plan; and if a vertical joint were allowed, C K would be the joint line for the mould, which is exactly similar to one produced by the system of ordinates, the plank having the same pitch. But it is neither allowed nor desirable, the butt joint being the more easily made, and looking by far the best when executed.

᛫he direction for the longer diameter in this and all similar cases will be parallel to A C through K, and the shorter axis at right angles to it in the line K B. This is owing to both pitches of the wreath being the same. The lengths of half the shorter diameters for both inside and outside curves will be equal to the radius for the inside and outside of the rail on plan. To find half the longer axis for the inside curve, take half the shorter one in the compasses, and with one foot at o draw an arc cutting the longer axis at Q. Through Q draw O Q R, meeting B K in R, and O R will be half the longer diameter sought. The major axis of the outer curve is found in the same way from P or N.

Now draw the quarter ellipses in the manner before described. Then draw the joint through c square to B C, and from the point where this joint line is cut by the convex curve, set off the whole width for this end of the mould towards the concave rather full, and continue the inside curve to this point, and the mould will be complete.

Fig. 4 shows the method of drawing the upper mould. C E is made equal to H E (Fig. 2), C D and D E are alike in both figures, and drawn as before. E L and C L are parallel to C D and D E. Then L is the centre of the elliptic curve. The distance

from E to the end of the mould is made equal to
the distance from E to the joint (Fig. 2).

To find the bevel and width for the top end E,
draw c c at right angles to D E, and with the
distance c c as a radius and the point 9 (Fig. 1)
as a centre, draw an arc cutting the centre of rail
at 10.   Draw 9 10, and this will be the required
bevel.   Parallel to 9 10, set off half the width of
the rail, and 10 11 will be half the width of the
mould.   Repeat this distance on each side of D E
(Fig. 4), and draw parallel lines as before.

For the lower end c, draw E e at right angles to
D c, and with E e as a radius and the point 12 (Fig.
1) as a centre, draw an arc cutting the centre line
of rail at 13 ; then the line 12 13 will be the bevel
required.   Parallel to it set off half the width of the
rail, and 13 14 will be half the width of the mould.
Repeat this distance on each side of c D (Fig. 4),
and draw parallels to cut c L ; then the points
where these parallels to c D cut c L will be those,
together with the points on E L, from which to find
the direction and length of the diameter.   These
are obtained and the curves described as on Plate
III.   Draw the joint through c square to c D, and
from the point where this line is cut by the convex
curve, mark off the whole width towards the
concave, and continue the inside curve to this
point.

The remarks which we made as to the quarter ellipses of the lower mould apply also to this upper one. Thus, for a vertical joint, c l would be the joint line, and if this mould were placed to its pitch over the plan, the line c l would lie in the same direction as the centre line of the well-hole, with c perpendicularly over 3, and the two points in c l denoting the inside and outside of the mould, perpendicularly over the inside and outside of the rail on the joint line of the plan.

The moulds are exactly the same as if produced from ordinates, the plank of course having the same pitch; and anybody who understands how to get them out by ordinates, will know that they must be correct. It will perhaps be noticed that the piece added to the inside of the lower mould is greater than that added to the inside of the upper one. The cause of this is the difference in the pitch of the two wreaths, the top one being flatter than the bottom one. So in every case the sharper the pitch the greater becomes the piece to be added inside to make the joint a square butt joint, and *vice versâ*, until it happens that either of the lines A B or B C becomes level, as in Plates II., V., and VII., when there will be nothing to be added, and the joint line will always be at right angles to such level line, and also in the direction of the diameter of the ellipse.

D

11. Those who may wish to satisfy themselves
further upon the above point, cannot do better
than get two pieces of wood one inch square.
This will be large enough to include the plan of
the rail. On the sides 1 2 and 2 3 (Fig. 1) of
the first piece, mark the pitches A B and B C (Fig.
2). Cut off to this bevel. On the sides 3 4 and 4 5
(Fig. 1) of the second, mark the pitches C D and
D E (Fig. 2), and cut off to this bevel. Draw the
line corresponding to tangent 2 3 and to B C of
the mould on the bevelled end of the first piece ;
also the line corresponding to tangent 3 4 and
C D of the mould on the bevelled end of the second
piece. Then screw the two together, having the
joint on the centre line of the well-hole, and
keeping the end of the lines on the bevelled ends
of the pieces corresponding to C of the moulds to
the same height. On the square end of the two
pieces thus screwed together, draw the half
circular plan of the rail. Take off the stuff to
these lines, leaving it concave on the inside and
convex on the outside, corresponding to the inside

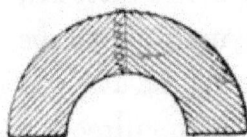

Fig. s.

and outside of the rail on plan.
The plan and elevation will
then appear as in Figs. s and
T, Fig. s being the plan, and
Fig. T an elevation of the con-
cave side. From this it will be evident, first,

that the quarter ellipse of the moulds must
finish at the points we have indicated ; secondly,
that these points will always be perpendicularly
over the inside and outside of the rail on the joint
line of the plan ; thirdly, that the triangular piece
we added to the inside of
the bottom mould belongs
more properly to the top, and
the piece we added to the
inside of the top belongs more
properly to the bottom ; that
is, if we divide the concave
side of the rail plan into two
equal parts, and from the
point of division erect a perpendicular across the joint of the
wreaths when together, each
of the pieces we have added

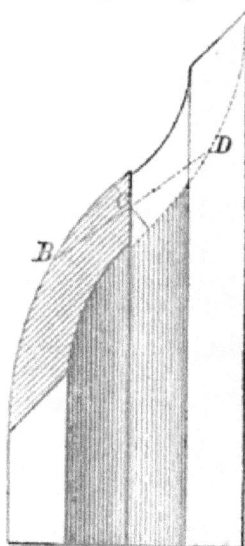

Fig. T.

will be found to belong to and form part of the
opposite quarter circle. The concave side of the
mould, therefore, when the joint line is drawn,
covers more than a quarter of a circle. But we
cannot draw more than a quarter of an ellipse on
the one mould when the other has a different pitch.
We therefore draw the quarter ellipses first, then
the joint line square to the tangent, and continue
the inside curve to the joint line as directed.

It will be evident from these considerations that

to draw the joint square to the tangent, and to set
off half the width of the mould on this joint line
on each side of the tangent, will be incorrect. To
draw the curve of the ellipse with any accuracy
through points so obtained will be found im-
possible, especially when the wreath has a very
sharp pitch. The centre point will be right, but
that towards the convex side will be too far from
the centre, and that towards the concave will not
be far enough.

This will appear more plainly from an examination of Fig. v. Let the two semicircles represent
the plan of the rail. The
line dividing it into two A
quadrants and drawn to the
centre will be the joint line
of the plan, and A E will be
the tangent for the centre of the mould. When this
is placed to its pitch over the plan, the tangent of
the mould will lie in the same direction as A E, and
the centre will lie perpendicularly over E of the
plan.

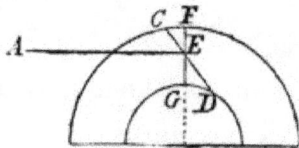

Fig. v.

But the joint line of the mould, instead of
being over the joint line of the plan, will be
more in the direction of C E D. (This may easily
be proved by drawing the joint through C square
to B C on the bevelled end of Fig. T.) Suppose
we find the width of the mould here to be 5¼

inches. If we set off $2\frac{3}{4}$ inches each side of the point perpendicularly over E, it will be evident that neither of the points so obtained will be perpendicularly over either outside or inside of the rail on plan. That towards the convex side will fall without the rail, and that towards the concave will fall within it. But instead of this, find the points perpendicularly over F and G. Draw the quarter ellipses to those points, and complete the mould as directed.

The application of these moulds to the plank, the jointing up and the bevel for same, will be similar in all respects to those described in the preceding Plates.

Make the casings, or square up the wreaths roughly, but do not finish them until the joints have been made and the parts bolted together.

PLATE V.

WREATHS FOR HALF-SPACE OF WINDERS.

---

12. This plate exhibits the moulds of a hand-rail for a half-space of winders, with a straight flight below, and the floor or landing at top.

Fig. 1 is the plan with the riser lines laid down as shown, and tangents drawn to the semicircular centre line of the handrail.

Fig. 2 is the development of Fig. 1 in elevation, the vertical lines 1, 2, 3, 4, 5, coinciding with the points 1, 2, 3, 4, 5, of the plan, and the riser lines drawn as they occur on the centre line of the rail, and the tangents to the semicircle.

The under side of the level rail at top is drawn at the distance of four inches from the floor, and the centre at the distance of half the thickness from the under side. This is continued in a straight direction, to cut the vertical line 4 in D. From D continue the centre downwards to meet the centre of the lower straight rail, so as to form a fair easing on the same. Mark the joints as shown, square to the falling line. The

PLATE V.

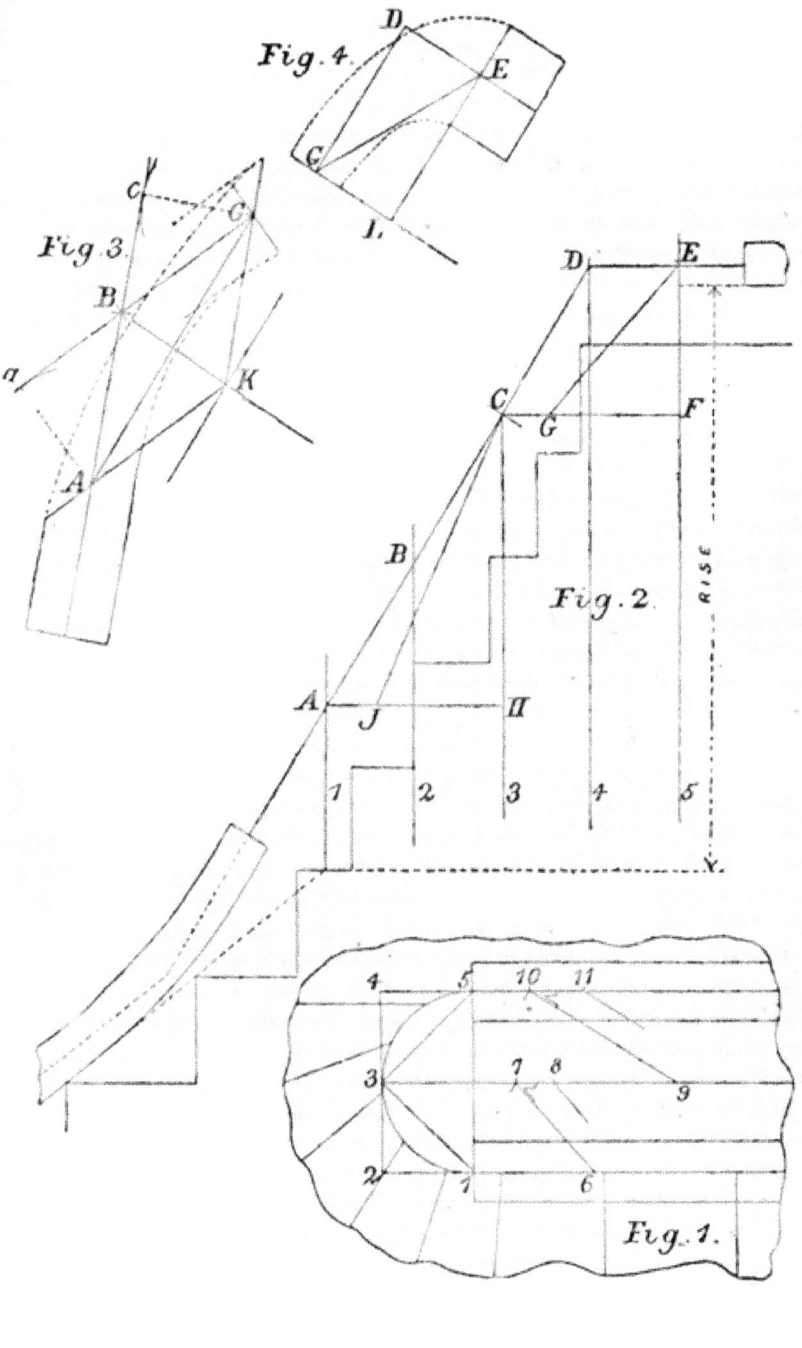
Fig. 4
D
E
G
L
Fig 3.
C
G
B
a
K
A
Fig. 2
D
E
C
F
G
B
RISE
A
J
H
1
2
3
4
5
Fig. 1.
4
5
10
11
3
7
8
9
2
1
6

rise or height for jointing the wreaths to the lower rail is obtained as on the preceding Plates.

From A and c draw the horizontal lines A H and c F. From H and F set off the distance H J and F G, equal respectively to the diagonals, 1 3 and 3 5 (Fig. 1). Draw J c and G E. Then J c will be the diagonal for the lower mould, and G E the diagonal for the upper one.

Fig. 3 shows how to construct the lower mould. Make A c equal to J c (Fig. 2). Draw A B and B c as before, and draw parallel to them respectively c K and A K. Then K becomes the centre for the elliptic curves. Both pitches of this wreath being alike, the longer diameter will be parallel to A c, through the centre K, and the shorter one through K at right angles to the longer one.

From A to the end of the mould is made equal to the distance from A to the joint (Fig. 2). The bevel and width for each end is determined as before, by drawing A a at right angles to B c, and the distance A a is then applied as at 6 7 (Fig. 1) for the bevel, and half the width of the rail drawn parallel to 6 7, in order to obtain 7 8, which is half the width of the mould. These both apply to each end of the mould. The distance 7 8 is drawn on each side of, and parallel to, A B (Fig. 3), to cut the springing A K; and where the same parallel distances to B C cut C K will be the points (together with those in A K) to which the quarter ellipses should be drawn.

Find the direction and length of the longer diameter, and draw the curves of the mould to those points as before. Then draw the joint square to B C, and from the point where this joint line is cut by the convex curve, mark off the whole width towards the concave side, and continue the inside curve to this point. The mould will then be completed.

Fig. 4 shows the method of constructing the upper mould. This presents no difficulty, the tangent D E being level. We proceed, however, by the same method as before, and make C E equal to G E (Fig. 2). We then draw C D and D E, and their parallels E L and C L, and L will be the centre of the curve. From L to C will be half the shorter diameter to the centre of this end of the mould (equal to radius for centre of rail on plan), and from L to E will be half the longer diameter to the centre of the top end.

To find the bevel and width for the top end, take L E or C D, and apply it from 9 to 10 (Fig. 1). This gives the bevel. Set off from 9 10 half the width of the rail, and 10 11 will be half the width of the mould. Draw parallel lines at this distance to D R (Fig. 4), to cut the springing line

L E. Make the distance from E to the end of the mould equal to the distance from E to the joint (Fig. 2).

The lower end requires no bevel. L c, being the shorter diameter, will be level, and the narrowest part of the mould is always on this diameter. The width here will therefore equal the width of the rail on plan. Set off half this width on each side of c on c L; then we have the distance from the centre L to the inside and outside of the mould on each diameter for drawing the curves.

When cutting this piece out of the plank the stuff should be left full at the narrow end.

Owing to the sharp pitch of this wreath, the mould produced at the top end has a great width, and if the piece were cut out to the full width of the mould, there would be more stuff used than necessary.

By the use of the bevel for this end the necessary width may be ascertained to a nicety. Thus, let A B C D (Fig. W) represent the top end of the piece, as it would be if cut out square to the full width of the mould E F G ($7\frac{1}{2}$ inches). Through the centre o of the width and thickness draw the bevel B D (the same as found at 9, 10, 11, Fig. 1, Plate V.). On each side of this line set off half the width of the rail, and draw the parallels to B D. On the bevel line B D set off half the thickness of the rail on each side of the centre o.

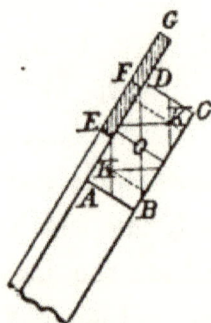

Fig. w.

Draw the top and bottom, and through the angles K K draw the dotted lines square to the face of the piece. Then these dotted lines give the necessary width of stuff to produce the wreath. Thus, instead of $7\frac{1}{2}$ inches, only 5 inches are needed, which is only one inch wider than the rail. The mould, however, must be retained the full width for bevelling the wreath, which is done in the same way as any other of a similar construction.

13. A word or two as to the lengths of balusters that are necessary here, and on some of the other Plates, may not be considered out of place. Certainly the under sides of the wreaths are not drawn so that the balusters on the winders may be the same length as those on the flyers. To have done so would have spoiled the appearance and crippled the falling line altogether, besides making it more dangerous for persons using the stairs.

It is usual with some joiners to make an easing on the upper straight rail as well as on the lower one, and to make the under side of the wreaths lie close to the narrow ends of the winders, so that the rail may be the same height and the balusters have the same length as on the flyers. But this does not appear to us to be the safest plan, nor do we think that the rail always looks the best when executed in this way. It is not worth while to cripple the falling line of the wreaths and make them look very unsightly, as they sometimes do, for the sake of getting the balusters all the same length. It is better to execute the wreaths and easing so that the two combined shall have a good even-looking falling line, and to let the lengths of the balusters take care of themselves. These are, however, questions which every one must be left to settle for himself according to his taste and judgment. But we think the rail has a much better appearance when lifted over winders, especially if the ascent is sharp, as in small well-holes.

It will be noticed that in no single instance have we drawn an easing on the upper rail in any of the Plates described in these pages. Nor should we ever think of doing so unless compelled by circumstances which we could not control, or by superior authority.

This we believe to be the correct system, if one may be said to be more correct than another; and this we think will produce by far the best-looking handrail when finished.

But whichever way this question may be decided, it will in no way interfere with the method of getting out the moulds as taught here. The one thing necessary is to see that the centre falling line is leading in the same direction on each side of the joint.

PLATE VI.

———

14. This Plate shows the construction of a mould for an obtuse angle with winders, having straight flights above and below. The general method of the preceding Plates is followed here in every particular, except in finding the centre for the elliptic curves of the mould.

Fig. 1 is the plan with the tangents 1, 2, 3 enclosing the circular centre of the rail. The radius lines 1 D and 3 D are drawn square to the straight rail from the centre D. The diagonal 1 3 is drawn between the two ends of the circular arc; D 2 will then be at right angles to the diagonal 1 3.

Fig. 2 represents the unfolding of the centre line of Fig. 1, the riser lines being placed as they occur on the centre, and the tangents. The vertical lines 1, 2, 3 of Fig. 2 will then coincide with the points 1, 2, 3 of Fig. 1.

Draw the falling lines of the rail as before, and continue the upper centre in a straight line down to B, and from B downwards, so as to get a good easing on the lower straight rail.

From A draw the horizontal line A E, and from E set off E F, equal to the diagonal 1 3 (Fig. 1). Draw F C; then F C will be the diagonal for the mould.

Fig. 3 shows how to draw the mould. Make A C equal to F C (Fig. 2). Then draw A B and B C as before, extending each line as shown, and making the distances from A and C to the ends of the mould equal to the distances from A and C to the joints (Fig. 2). Draw the ends of the mould square to A B and B C.

To find the centre of the elliptic curve, bisect A C in G. From B draw B G. Take B G in the compasses, and with one foot at 2 (Fig. 1) draw the arc cutting the diagonal 1 3 at H. From 2 draw the line 2 H, extending it to meet the perpendicular on the line D 2 from D at J. Make B G D (Fig. 3) equal to 2 H J (Fig. 1). Then D will be the centre of the curve sought.

From D draw D A and D C through A and C. These will be the springing lines, and will denote on each outside edge of the mould the point where the straight and circular parts will meet.

The bevels and widths are obtained as on any of the foregoing Plates.

Thus, for the top end, draw A a at right angles

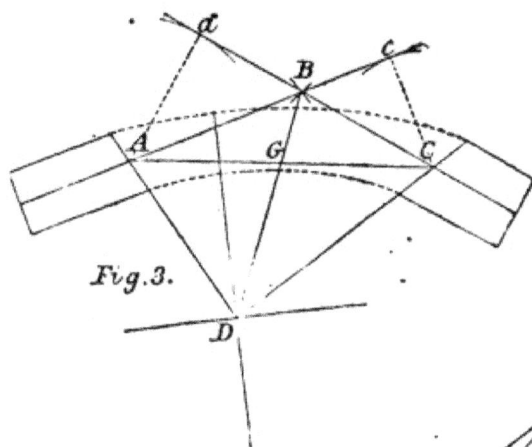

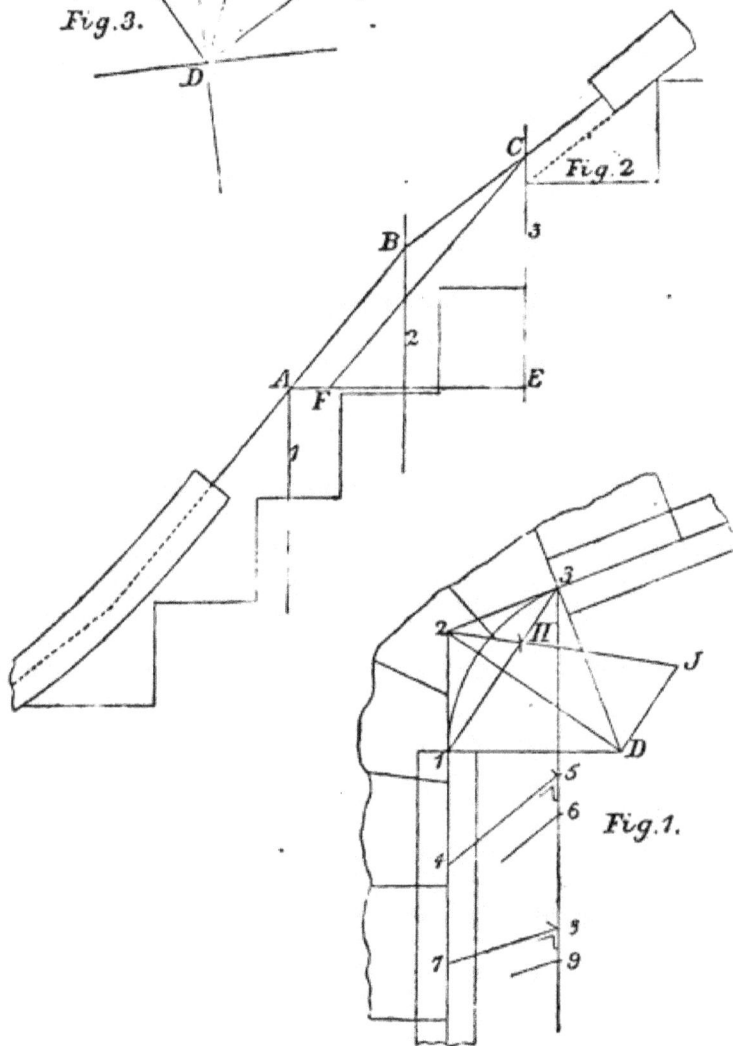

PLATE VI.

Fig.3.

Fig.2.

Fig.1.

to B c. Then, with the distance A a for radius, and
the point 4 (Fig. 1) (anywhere on centre line
of rail) for centre, draw the arc cutting the
centre line at 5.   Draw the line 4 5, and this
will be the required bevel.   Parallel to 4 5 set
off half the width of the rail, and 5 6 will be half
the width of the mould.   Repeat this distance on
each side of B c (Fig. 3), and draw parallel lines
to cut the springing line D c.

For the bottom end the same method is followed, and the bevel and width, 7, 8, 9 (Fig. 1),
obtained as above.   This again is repeated on
each side of A B (Fig. 3), and parallel lines drawn
to cut the springing line A D.

Thus we have the two points in each edge of
the mould, the centre D, and the length of half
the shorter diameter, which will be equal to the
radius of the plan for the inside and outside of
the rail.   From these we find the direction and
length of the diameter, and draw the curves in
the same way as on any of the foregoing Plates.

The application of this mould for the purpose
of bevelling the wreath is the same as for a right-
angled plan.

The method of proceeding with an acute angle
(when the diagonals are at right angles to each
other) will be the same as for an obtuse angle. The
one Plate will therefore serve for the two angles.

PLATE VII.

MOULDS FOR SCROLL SHANKS.

---

15. This plate shows the application of the author's method of handrailing to the production of moulds for scroll shanks. A very brief description only will be necessary, as the mould is drawn in a manner exactly similar to that used on Plate II.

Fig. 1 is the plan, having the centre of the largest quadrant of the scroll enclosed, with the tangents 1, 2, 3. The diagonal is drawn from 1 to 3, but it might be dispensed with in this and all similar cases, there being no difficulty in obtaining the angle B (Fig. 3), because it is a right angle.

Fig. 2 is the development or elevation on the centre line and tangents of Fig. 1, showing the top of scroll step and the next flyer. The under side of the rail is drawn resting on the angles of the flyers, and the centre line at the distance of half the thickness. Prolong the centre of the rail downwards, to cut the vertical line 2 at B.

PLATE VII.

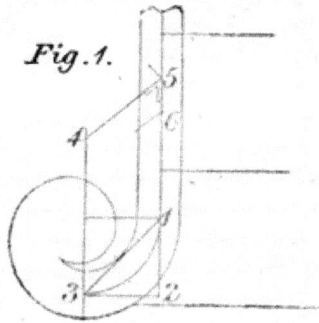

Fig. 1.

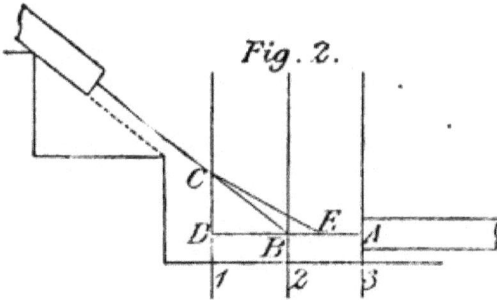

Fig. 2.

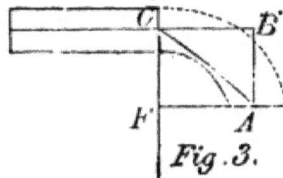

Fig. 3.

From B draw B A horizontal, and extend it to D. From D set off D E, equal to the diagonal 1 3 (Fig. 1). Draw E C, which will be the diagonal for the mould.

Fig. 3 shows how to draw the mould. Make the diagonal A C equal to E C (Fig. 2). Then take A B and B C (Fig. 2), and lay them down at Fig. 3, as before. Parallel to A B and B C draw C F and A F. The point F then becomes the centre of the quarter ellipses.

The only bevel we require here is one for the lower end. This is obtained by taking A F (Fig. 3) in the compasses, and applying it from 4 to 5 (Fig. 1). This will give the bevel. Parallel to 4 5, set off half the width of the rail, and 5 6 will be half the width of the mould. The top end will be square, and the same width as the rail on the plan. Set off the distances on each side of A and C, and draw the quarter ellipses of the mould as before.

The joint at 3 (Fig. 1) will be vertical, and may be made square from the under side of the level portion of the scroll, and square to the face and tangent A B of the wreath. It is better, however, to mark the springing on the inside edge of the wreath when it is bevelled, as in any other case. Then, in making the joint, place the level portion of the scroll on a level surface, and holding the

parts of the joint together, prove the springing
and the inside edge of the wreath with a set
square, both of which should be brought to a perpendicular position.

The upper joint presents no difficulty, the shank
end of the wreath being straight. In making it,
however, the centre of the wreath should be placed
to the centre of the rail, except when it requires
raising or lowering to bring the height of the
newel right.

This method of drawing the mould in cases of
this kind might be dispensed with by using
ordinates and the pitchboard, which would give
the same bevel and pitch of plank. If, however,
the tangent A B (Fig. 2) were drawn out of a
level, the pitchboard would not give the bevel
nor the pitch of the plank ; and the joint for the
level portion of the scroll, instead of being square
to the under side, would have to be bevelled. So
that if by drawing A B horizontal the length of
the newel becomes too great, we have only to
decide what length it shall be, and draw the under
side of the scroll to suit this height. Then at
the distance of half the thickness draw the centre,
and connect this with the point B, letting the
centre of the joint still be on the vertical line 3,
but taking care that the line from B to meet the
level centre shall be continued in a straight direc-

tion some little distance beyond the point A. This, perhaps, will be seen more clearly by referring to the following Plate, where we have been obliged to enter more minutely into the matter.

PLATE VIII.

———

16. This Plate gives an example of the construction of a mould for a scroll shank with commode steps, or diminished flyers, at bottom. In this case there are two steps only. The number, however, may be either more or less, according as different situations may require them.

The method of proceeding will be the same as that adopted in the foregoing Plates, except in the mode of finding the centre for the ellipse, which will be explained in its proper place.

Fig. 1 is the plan with the risers laid down as shown. Draw the tangent 1 2, which is a continuation of the centre line of the rail. Also draw the joint at 3, radiating towards the centre from which this part of the scroll is described. At right angles to this joint line, and from the centre 3, draw the tangent 2 3. Next draw the diagonal from 1 to 3; then from 2 draw 2 4, pointing to the centre from which the largest quadrant of the scroll is struck.

PLATE VIII.

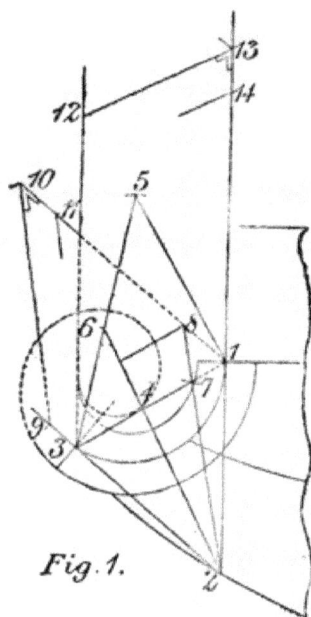

Fig. 1.

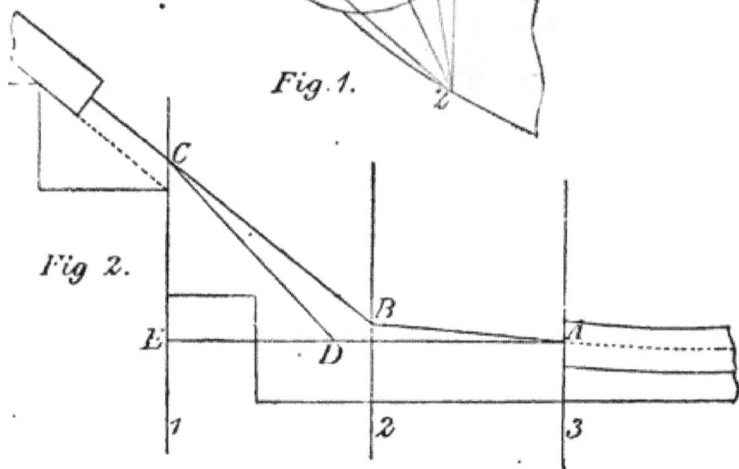

Fig 2.

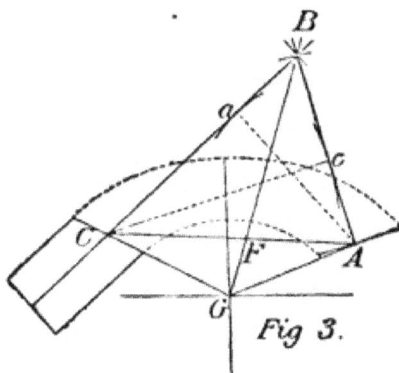

Fig 3.

E 2

Fig. 2 unfolds in elevation the centre line and tangents of Fig. 1, the risers being placed in the development as they occur in the plan, on the centre line, and tangents. The distances 1 2 and 2 3 (Fig. 2) will then be equal to 1 2 and 2 3 (Fig. 1), with the flyers at top, the curtail step at bottom, and the diminished flyer between them.

Draw the under side of the straight rail resting on the flyers, and at half the thickness of the rail draw the centre line. Continue this in a straight direction, to cut the vertical line 2 at B.

In order to decide upon the height of the newel, suppose the top of the rail is to be 2 feet 8 inches in vertical height above the tread over the front of the flyer riser line. This rail will measure 4 inches in vertical thickness, and will therefore require lifting 2 feet 4 inches. If, therefore, the newel is to be 2 feet 6 inches, the under side of the level portion of the scroll will have to be drawn at a distance of 2 inches above the top of the curtail step. Draw the under side of the scroll at this distance, and the centre line at half the thickness of the rail. Then from B draw B A, to meet the centre of the level scroll some little distance beyond the vertical line 3 A. Draw the lower joint through A square to A B, and the upper one at any convenient distance from C. From A draw the horizontal line A E, and from E

set off E D equal to the diagonal 1 3 (Fig. 1).
Draw D C, which will be the diagonal for the
mould.

Fig. 3 shows how to draw the mould.   Make
A C equal to D C (Fig. 2).   Then A B and B C (Fig. 3)
are respectively equal to A B and B C (Fig. 2), and
a.e drawn as before described.   A C, the diagonal of
the mould, is the line that wi'l lie over the diagonal
1 3 (Fig. 1) from end to end in an inclined position.   Take therefore A C (Fig. 3) in the compasses,
and with one foot at 3 (Fig. 1) draw an arc, cutting the perpendicular to 1 3, at 5.   Draw 3 5,
and from 4 draw 4 6 parallel to 1 5 ; then 5 6
will be the distance from C (Fig. 3), through which
to draw B F.   Therefore make C F (Fig. 3) equal
to 5 6 (Fig. 1), and draw B F.   To find the
length of B G, take B F in the compasses, and with
one foot at 2 (Fig. 1) draw an arc cutting the
diagonal 1 3 at 7.   Draw 2 7 and extend it to 8,
where it cuts the line parallel to 1 3, drawn from
the centre of the largest quadrant of the scroll.
Then B G (Fig. 3) should be made equal to 2 8
(Fig. 1), and G will be the centre of the elliptic
curves.

The length B G may be determined on Fig. 1 at
either side of the point 4, whichever may be found
most convenient, as either will be correct.

This method of finding the length of B G it

necessary, owing to 2 4 not being square to 1 3 (Fig. 1), and it can be drawn square only when the centre of the largest quadrant of the scroll is in such a position as to admit of it. This of course will depend upon what direction the diagonal is drawn in. If it could be so drawn that a line at right angles to it from 2 would pass through the centre of the largest quadrant of the scroll, then the point F in A C (Fig. 3) would be in the middle of A C, and the length B G might be found in the same way as in the case of an obtuse or acute angle.

Having found the centre G of the ellipses, draw the lines A G and C G.

The bevels and widths are found according to the general method given above. Thus for the bottom end draw c c at right angles to A B (Fig. 3). Then at Fig. 1 extend the tangent 2 3 to 9, and from 1 draw 10 11 parallel to 2 3. Then, with the distance c c as a radius, and the point 9 for a centre, draw an arc cutting 1 10 at 10. Draw 9 10, and this will give the bevel. Parallel to 9 10 set off half the width of the rail, and 10 11 will be half the width of the mould. Repeat this distance on each side of A B (Fig. 3), and draw parallels to cut A G. For the top end, draw A a at right angles to B C (Fig. 3). Then from 3 (Fig. 1) draw 3 12 parallel to 1 2; and with A a

for a radius and the point 12 for a centre, draw
the arc cutting 2 1 produced at 13. Draw 12
13, and this will be the bevel required. Parallel
to 12 13 set off half the width of the rail, and
13 14 will be half the width of the mould.
Repeat this distance on each side of B C (Fig. 3),
and draw the parallel lines to cut the springing
line C G.

Thus we have two points in the inside curve,
the centre G, and half the shorter diameter of
the ellipse, which last will be equal to the radius
for the inside of the largest quadrant of the
scroll at Fig. 1. From these we determine the
direction and length of the longer diameter,
and can draw the curves in the same manner
as before. After describing the curves, draw
the joint at A square to A B, and the mould will
be completed.

We have drawn the lines at Fig. 1 for obtaining the bevels beyond the scroll for the sake of
clearness. This, however, is not necessary.
Having taken the distance at Fig. 3 for the top
end, one foot of the compasses may be placed
at 3 (Fig. 1), and the arc drawn to cut the line
1 2 in the same way as we have drawn it. And
for the bottom end, with the distance taken at
Fig. 3 in the compasses, one foot may be placed
at 1 (Fig. 1), and the arc drawn to cut the

tangent 2 3. Either way will be correct, the one method producing the same bevel and width as the other.

This wreath is bevelled in exactly the same way as any other with two pitches.

The joint at the upper end may be made at any convenient distance from c, as before stated, and square to the face and inside edge. Let the centre of the wreath be placed to the centre of the rail, unless it should require slightly lowering or raising, in order to bring the under side of the level piece to the proper height of the newel.

For the lower joint, a bevel will have to be set with its stock to the under side of the level portion of the scroll, and the blade made to coincide with the joint line drawn through A.

The bottom end of the wreath may be planed square to the face, and to the tangent A B. It is much better, however, to make this lower joint as advised on Plate VII. Thus, plane the joint on the level portion of the scroll to its proper bevel and direction across the face. Then fix this level piece to a flat surface, and fit the lower end of the wreath until the springing line and inside edge are proved by a set square to be in a perpendicular position.

PLATE IX.

HANDRAIL FOR STAIRS WITH WINDERS AT
BOTTOM.

———

17. This Plate exhibits a plan of stairs and rail that might possibly occasion some difficulty to any one not very well versed in the foregoing method of getting out the moulds for wreaths, and coming upon a plan of this kind for the first time. The general method, however, already given for finding the direction and length of the elliptical axes for the mould will apply in exactly the same way here as in former cases. The only part of the plan that should cause any difficulty is the construction of the mould for the upper wreath, Fig. 4.

This is no fancy plan, but one that not unfrequently occurs in actual · practice, the author having met with such; and he has also been asked for information as to how to proceed generally with such a plan, and more particularly how to get out the moulds.

Fig. 1 is the plan of a staircase with a curtail

PLATE IX.

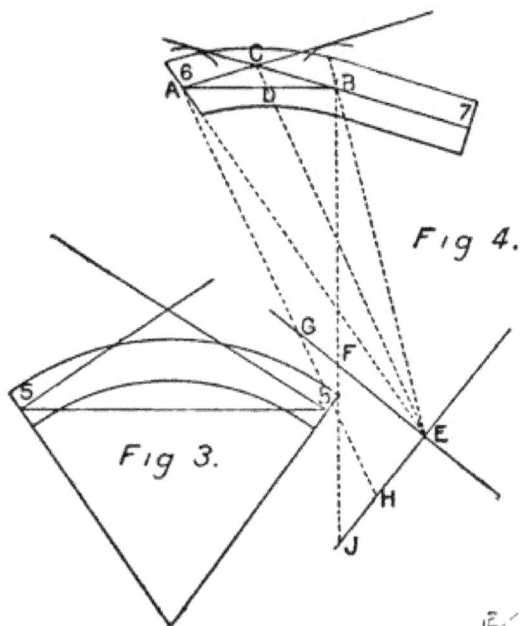

Fig 4.

Fig 3.

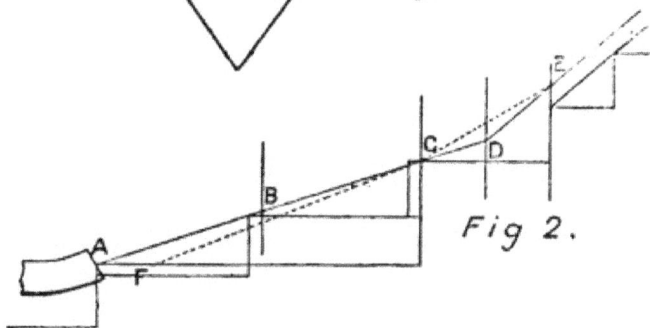

Fig 2.

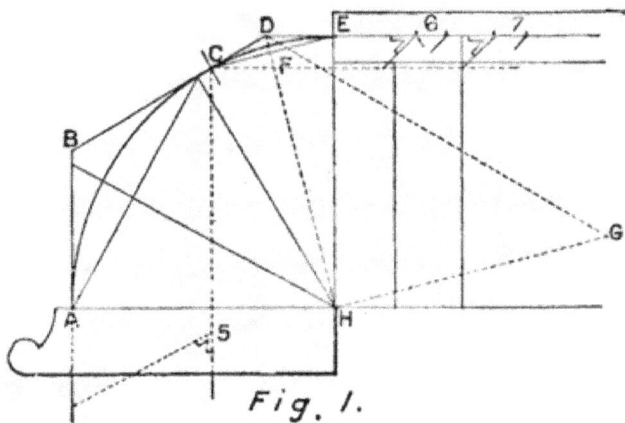

Fig. 1.

step at starting and winders in the quarter space, the narrow ends of which may be finished against a newel, and the right-hand end of the flyers housed into a wall-string. The centre line only of the quadrant is laid down, and this is enclosed by tangents, as shown at A B C D E. The joint connecting the bottom end of the lower wreath to the scroll should be on the line where the curves running in opposite directions meet. In Fig. 1 it is shown at A on the second riser line. It may be placed here or moved either way as required, provided the above requirement is not forgotten.

The tangent D E might be made shorter if it should be thought desirable, thus bringing the point D nearer to E, and thereby increasing somewhat the length of A B and B C. The effect of this would be to make the point D in the development (Fig. 2) somewhat higher than is there shown, and thus lift the upper end of the first wreath, which would perhaps be an improvement.

Fig. 2 is the development of the riser lines and tangents of Fig. 1. The first wreath, as will be observed, lies very flat, and the small easing required to bring it from the rake to the level of the scroll eye may easily be made by having the piece of stuff for the scroll extra thick. The bottom end of this wreath, as at A, may be lifted or lowered as may be considered most desirable.

The diagonals F C and C E (Fig. 2) for the moulds are obtained as already fully described.

The mould for the first wreath (Fig. 3) is constructed in accordance with instructions previously given, and need not be further described here.

The mould for the second or upper wreath (Fig. 4) is one that may require some explanation, although it is constructed in a similar way to the lower one. Make the diagonal A B equal to C E (Fig. 2), and the tangents A C and B C each equal to C D and D E (Fig. 2). Bisect A B in D, and draw C D, producing it indefinitely. Take C D in the compasses, and with one foot at D (Fig. 1) describe an arc cutting C E in F. Through F draw D F, and produce it to meet the perpendicular to D H in G. Then in Fig. 4 make C E equal to D G (Fig. 1), and E will be the centre of the elliptic curves for the mould.

To find the directions of the axes apply the square as in former examples, with the end of the line in the centre of it placed to E, as shown. Then take the radius to centre of rail in Fig. 1, and apply it from A and B (Fig. 4) to cut the end of the square in F and G; and when lines drawn through these last points from A and B, and produced to H and J become equal, each to each, the end of the square will give the direction of the axes as before, and the mould may be described as already explained.

It will thus be seen that the only difference between this and all preceding examples of a similar description is that, while in the latter the two points A and B are situated one on each contiguous side of the minor axis, they are in this case both on one and the same side of it. This circumstance, however, as will be evident from the example before us, makes no difference whatever in the method of procedure.

The situation of the two points A and B in the circumference of the ellipse, of which the curves of the mould form a part, depend of course upon the position into which the plank is thrown for cutting out the wreath, and also what part of the circumference the mould curves refer to, the former condition always determining the latter.

As a rule, the mould is cut, or crossed, by the minor axis of the ellipse, or, in other words, the curves of the mould consist of a portion of two elliptical quadrants which are situated one on each contiguous side of the minor axis. When this is the case—and it is almost always—the place where the mould is crossed, or cut, by the minor axis is perfectly horizontal when placed in its true position, and the same width as the rail is on plan.

In the case before us (Plate IX.) when the mould (Fig. 4) is placed in its true position, that is, the position which the plank occupies for cuting out the wreath, no part of it is in a horizontal position, and it is everywhere wider than the rail on plan.

The same class of moulds is produced in using other systems of lines on Handrailing.

SQUARING THE WREATH.

18. As this perhaps has not been made sufficiently clear in the foregoing pages, we will endeavour to state the method we adopt as fully and explicitly as possible.

Take such a pair of wreaths as those shown on Plate II., where the straight rail has no easing. Having bevelled them according to directions, the easing may be roughly made as before described. That is, some of the superfluous wood may be taken off at top and bottom, or in other words they may be roughly squared. But the stuff should be left very full all round, as it is difficult to tell where to take off and where to leave on to a nicety with a single wreath until it has been attached to the straight rail.

The joints being made as directed, take the bottom wreath and bolt it to the rail. See that the straight end of the inside, or concave edge, is in a line with the edge of, and square to, the under side of the rail. Clean off the outside edge of the wreath to width of rail. Then take off the top and bottom to a distance of 4 or 5 inches square with the edge, and in a line with the top and bottom of the rail. Go through the same

operations with the upper end of the top wreath
and rail. Then take them off the rail, and bolt
the parts of the centre joint together. See that
the springing lines are true, or out of winding
one with the other. Clean off the joint on the
concave side, so that the half circle may be seen
without any cripples by looking on the top.
Gauge to a width with a gauge like that shown
in Fig. z, slightly rounding the stem to fit the
concave side of the wreath,. and letting the
distance from the stem to the pencil point be
equal to the width of rail. Then apply
the stem to the concave side in a perpendicular direction, and run the pencil
round the convex top and bottom of
the wreaths. Take the stuff off to
these lines, to bring them to the proper
width all round. Make the easing on
the top side by taking off the super-   Fig. z.
fluous stuff between the two straight ends, so as
to produce an even-looking falling line, using a
square if it is thought necessary. Then gauge
to a thickness, and take off the stuff on the under
side, using a pair of callipers to see that the
proper thickness is maintained across the width,
as otherwise they are liable to be left thick in
the middle.

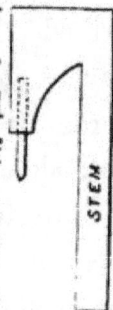

This will be found a safe and reliable method

to adopt, either with a pair of wreaths or with a single one.

Also when the lower rail has an easing the same method will apply. And if the wreaths have been jointed-up to their proper height, no unsightly easings and mistakes are likely to occur.

STAIRBUILDING.

STAIRBUILDING,

OR

THE CONSTRUCTION OF STAIRCASES.

———•———

PRELIMINARY.

1. IN treating of the construction of staircases we have not much to say that is not already well known to some at least of our readers, but our endeavour will be rather to explain the different processes to be gone through in constructing a staircase in as clear and intelligible a manner as possible, without omitting any matters upon which it may be advisable to offer instruction and advice to those who may feel the want of it.

The different kinds of staircases are pretty well known and understood, so far as their names are concerned, by all who are engaged in the building trade as carpenters and joiners. They may generally be divided into two classes—the dog-legged or newel stairs, and the continued or geometrical stairs. The former are mostly used for cottages and small houses, although some very elaborate specimens of this kind of staircase are occasionally met with in large and noble houses. In the case of cottages, however, the stairbuilder is sometimes at a loss to know what to do for the best, there being no room for anything worthy of being called staircase by which the occupants may get from floor to floor with ease and comfort. This of course can easily be remedied by having a proper plan, and making at least such provision for the stairs that they may be so constructed as to obviate any liability to accident by the persons using them. Staircases of the latter kind, that is, the geometrical, are as a rule used only in the better class of houses and mansions where elegance of form is desired, and in such cases the plans of the staircase are usually determined at the same time as the plans of the building, and suitable provision is made for them accordingly.

2. To make a staircase perfectly easy of ascent and descent the *going** and rise, or *tread* and *riser*, should always be properly proportioned to each other. The higher the rise the less should be the *going*, and the greater the *going* the less should be the rise. This proportion may be

---

* By the "going" is meant in all cases the distance from the face of any one riser to the face of the next one, and does not include the nosing or projection of the tread beyond the face of the riser.

arrived at in a very great number of cases by making the sum of the going and rise equal to from 16 to $17\frac{1}{2}$ inches, according as the *going* increases from 8 to 12 inches; the lower sum being increased by from about $\frac{1}{4}$ inch to $\frac{1}{3}$ inch for every 1 inch inc. .ase in the *going*, or as near this as the total *going* and rise of the stairs will allow.

Thus, if the *going* is 8 inches the rise may be the same, though it should not be more. If the *going* is 9 inches, the rise should be about $7\frac{1}{4}$ inches. With a *going* of 10 inches the rise may be $6\frac{1}{2}$ inches. If the *going* is 11 inches, the rise may be 6 inches. And if the tread has a *going* of 12 inches, the proper rise will be $5\frac{1}{2}$ inches. Either of these proportions may be used with the most satisfactory results, and the stairs will be found to meet every requirement for being ascended and descended with tolerable ease and comfort, besides which they will be found to embrace the vast majority of cases of varying rise and *going* likely to be met with in practice.

The number of risers, however, in a staircase belonging to the better class of work is, as a rule, decided by the architect, and from this of course there must be no departure without his authority.

But sometimes there is a little to spare in the total *going* of a staircase, and when this is the case the *going* of the separate treads may be increased or diminished somewhat, so as to make them approximate to the above proportions as nearly as may be considered desirable.

3. The Story Rod.

When the number of steps in a staircase is decided upon, the total height, or rise, is taken from floor to floor with a rod, called the "story rod," by standing one end of it on the floor below and marking it at its upper end by the height of the floor next above. It is then divided by a pair of compasses into the same number of equal parts as there are to be risers in the stairs, and the exact rise of each step will be equal to any one of those parts. Or the total height and the *approximate* rise only of each step being known, the number of risers and the *exact* rise of each may be obtained arithmetically thus—

Suppose the height for the stairs from the top of one floor to the top of the next above to be 11 feet 6 inches. Reducing this to its lowest denomination, we get 138 inches. Then let the required rise be about $6\frac{1}{2}$ inches, and divide the 138 by $6\frac{1}{2}$ and we obtain 21 as the total number of risers in the flight, and again dividing 138 by 21 we obtain $6\frac{4}{7}$ inches as the *exact* rise of each step.

CHAPTER I.

Plans of Staircases.

4. Before laying down the plan of a staircase, the first thing to ascertain is the size of the space the stairs are to occupy. We must also know the number of steps there are to be, and the rise from

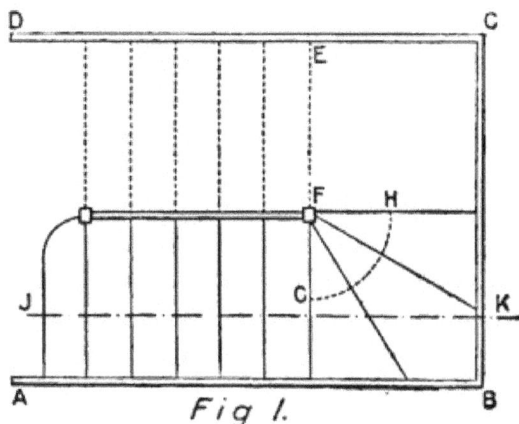

Fig 1.

floor to floor. These particulars being determined, the stairs may be delineated both in plan and section.

The first example is one of a dog-legged or newel staircase. Let A B C D, Fig. 1, be the plan of the walls within which the stairs are to be

erected ; the width, B C, being 5 feet 9 inches, and
the total height 9 feet 8 inches.   There being
16 risers in the plan before us, this will give a
rise for each step of $7\frac{1}{4}$ inches, for which a 9-inch
going would be a suitable proportion.

Draw first the lines representing the thickness
of the *wall-strings*,* allowing of course for the
plastering ; then the newel, F, the centre of which
should be on the riser line, E, which is drawn at
a distance from the back wall, B C, equal to the
semi-width of the staircase, and at right angles
to the side wall, A B.   With any radius describe
the quadrant G H from the centre of the newel F,
and divide it into the same number of equal parts
as there are to be winders (in this case three) ; and
through the points of division and from the *centre*
of the newel draw the lines representing the face
of the risers.

Draw the riser lines for the flyers at their
proper distance apart, in this case 9 inches ; also
draw the first newel and *middle-string*, and describe
the quadrant for the corner of the first step, and
the plan will be complete.   It may be remarked,
however, that the first newel in this plan is immediately under the last or top newel.   Conse-

---

* The term "string" is applied to the sloping plank into
which the ends of the treads and risers are *housed ;* the string
being cut out to receive the ends.

quently this latter cannot be shown by dotted lines as it should be if its position were on either side of the bottom one.

In making a plan of the above to work from, the winders and newel only, including of course the *strings* at the wide end of the winders, are laid down full size, the other parts being drawn to any scale from $\frac{1}{2}$ inch upwards, but a $1\frac{1}{2}$ inch scale, if not considered too large, will be found very convenient both in laying down the plan and also in working from, as every $\frac{1}{8}$ of an inch in such a scale will represent 1 inch, and every $\frac{1}{16}$ of an inch will represent $\frac{1}{2}$ inch.

By making a section (Fig. 2) of these stairs on the line J K, Fig. 1, they may be further delineated, and the length of strings, newels, and handrails may be ascertained preparatory to cutting off the stuff.

In Fig. 2 draw the line A B for a base, and set up the first run of steps to their proper rise, taking the *goings* from riser to riser as they occur on the line J K, Fig. 1. The first six being flyers will be equal to the *going* of the pitch-board, the seventh being a winder will be wider, and the eighth wider still.

From this point the figure will represent an elevation of the remaining part, inasmuch as it presents to view the outside face of the second string instead of the section of the steps as in the first run. The upper part is, however, set up in a similar way to the lower part, as shown by the dotted lines representing the ends of the steps, &c.

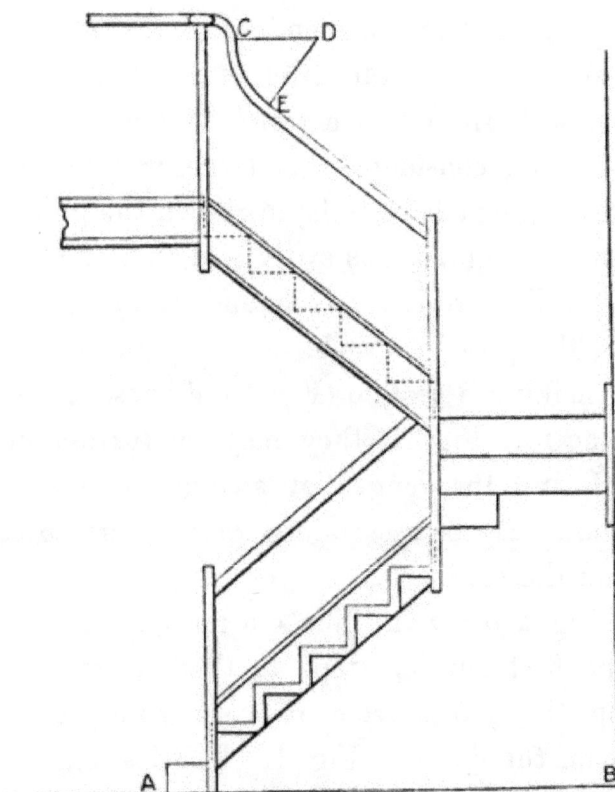

Fig 2.

If now the top edges of the *strings* are drawn parallel to and at their proper distance* from the

---

* This distance may be anything to suit the width of string, but a very general one is 2 inches or 2¼ inches from the angle made by the top of the tread with the face of the riser.

nosing-line* of the treads, and the width is then set off downwards, their lengths may be obtained very accurately. Then draw the handrails at their proper distance above the treads, and their lengths, and that also of the newels, may be measured off according to whatever scale the section has been drawn.

In fixing the floor joist for this and all other kinds of stairs care must be taken to see that the trimmers are placed at their proper distances from the walls, also the landings at their proper height, &c. To ensure having these matters attended to, it is advisable to lay down a plan of the stairs before the joists are put on at the building, and give instructions accordingly.

In Fig. 2 the handrails are shown without any easing at either end, and framed straight into the newel; the newel running up above the rail and having a turned or some other kind of ornamental top.

Sometimes there is a newel cap and an easing at both ends of the same piece of rail, the upper one, as shown by the curved lines at c e, being called a swan's neck. The method of finding the centre, D, from which the curves are described,

---

* The "nosing" is the amount of projection of the tread over the front of the riser.

will be sufficiently obvious from an examination
of the figure without a detailed description.

A better-looking easing, however, than can be
had by making it consist of arcs of circles, as in
the above, may usually be obtained by sketching
it freehand to suit the eye, and the bad appearance sometimes shown at the part c is thereby
avoided.

The easing at the lower end may be obtained
by the *pitch-board* as
shown in Fig. 3, where
A B C is the pitch-board,
with the under side of
the rail, F D, resting on
the hypothenuse, and D E
the under side of the
newel cap and horizontal
part of the easing, drawn
at about half the rise
above and parallel to the *going* A B. The curved
parts may then be described by finding their
centre as at G, or the angle may be eased off to
suit the eye.

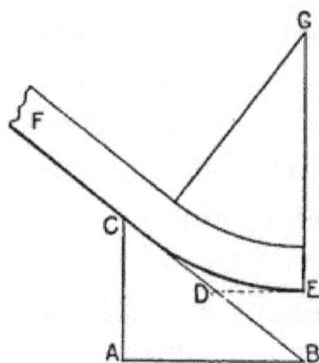

The upper edges of the wall-strings at top and
bottom of the stairs, and at the landings and
winders, are sometimes continued in a straight
direction to meet the top horizontal edge of the
skirting at an angle; but this is done only in the

commoner class of work. In a better class of work the top raking edges of the strings are brought to the horizontal by a gentle curved line to suit the depth of the skirting, so that the moulding or chamfer may be continued in an unbroken line from top to bottom. A good effect may be produced by this means if the curved easings are not too large and are carefully and evenly made.* That over the winders should be made suitable to whatever number there may be, and need not necessarily follow the line of nosings, so that the top edge may be at all points the same distance from them as it is at the flyers. Novices getting this idea into their heads will sometimes spoil the appearance of this part of their work, and produce a very unpleasing effect.

In a staircase with a plan like that shown in Fig. 1, the easing of the string containing the ninth and part of the eighth winders should be lined out first, then the upper end of the first long string can be eased round to meet the height of the former in the angle.

Fig. 4 shows the former easing, and Fig. 5 the upper end of the latter easing ; the height, A B,

* Various methods for drawing these curves are given in the author's book on "Circular Work in Carpentry and Joinery." Crosby Lockwood & Son.

above the eighth riser, or first winder, being the same in both figures.

The wide ends of the winders are, of course,

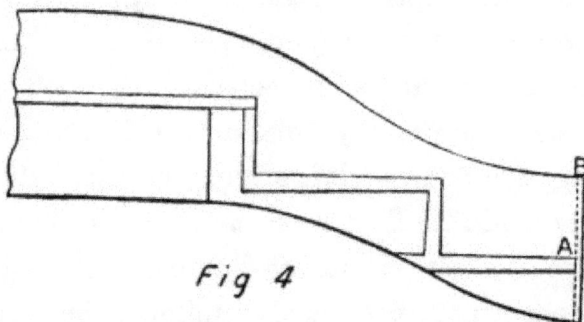

Fig 4

housed into the wall-strings in the same manner as the flyers, but the narrow ends and their risers instead of being housed similarly into the

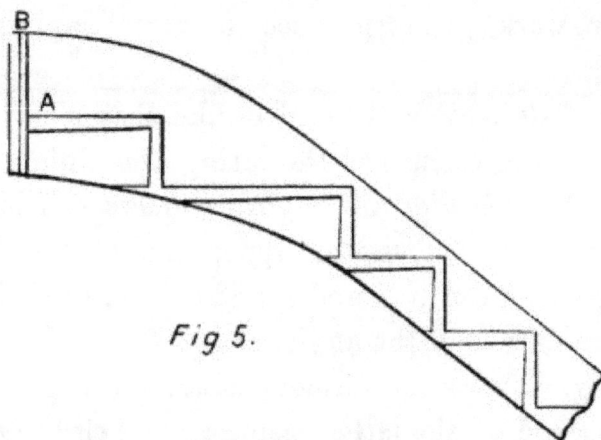

Fig 5.

newel are let into it from the back, the piece being cut clean out to the depth of about half an inch.

By this means the risers and winders being put into the housings of the wall-strings first, when they are being fixed, the opposite or narrow ends may be put into their places in the newel from the back without any difficulty.

The *outer-string* as shown in this stairs is what is termed a *close-string*, because the ends of the steps are housed into it in the same manner as into the *wall-strings*.

The top edge is usually finished with a capping to receive the balusters, and the lower edge with a return bead to finish fair with the plastering, the laths for which are sometimes nailed to the lower angles of the treads, if the strings are sufficiently strong to dispense with "carriages," or pieces of timber placed under their lower edges to stiffen them.

5. When the outer-string is cut to mitre with the ends of the risers, it is termed an *open* or *cut* and *mitred* string. The lower edge in this case is finished with a return bead or small moulding, and sometimes there is a sinking in addition to relieve the plain appearance it would otherwise present. With an open string, pieces of timber termed "carriages," and running in the same direction as the strings, will be required, on account of the strength of the strings, for efficiently supporting the outer ends of the steps,

being very much reduced by the cutting. The depth of the carriages will depend upon their length and the weight they may have to support. But whatever their depth the outer string must be wide enough to project sufficiently beyond the lower edge to include the thickness of the plastering.

The lower ends of the balusters with a cut

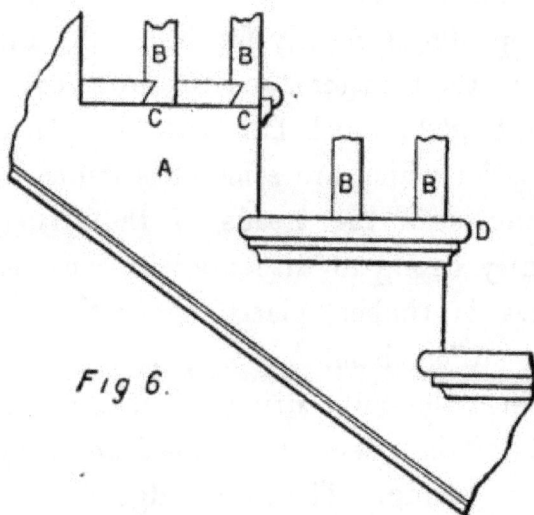

Fig 6.

string are dovetailed into the ends of the treads, and then hidden from view by a *return-nosing* which is moulded to the same section as the projection of the tread beyond the face of the riser, being mitred to it at one end, and returned in itself at the other.

Thus, in Fig. 6, A is the face of the string, B B the balusters dovetailed into the tread as at C on the upper step, and then hidden from view by the return nosing as at D, on the lower step.

For the circular corner of the first tread a piece of dry stuff the same width as the riser and thick enough to run beyond the springing at either side, is used, and cut to the proper shape.

The riser is then cut down to a thin veneer for a sufficient distance to go round the block, to which it is glued, screwed and wedged.

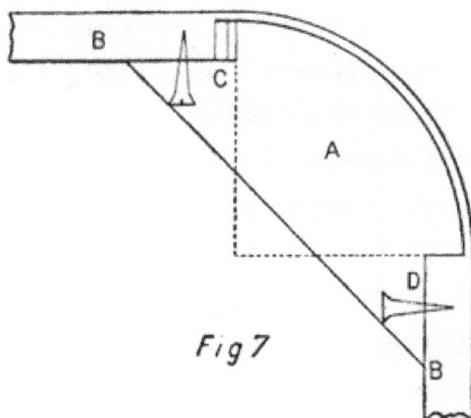

Fig 7

In Fig. 7, A is the block, which is first cut to the required shape, usually a quarter of a circle. It is then laid on the back of the riser and rolled round to get the correct length of the veneer, which will require to be about $\frac{1}{16}$th of an inch thick. The veneer being cut, the riser should be laid on the bench face downwards, and the end, D, of the block properly fitted and screwed. Then bend the veneer round the block gradually, by bringing its opposite end, c, down to the riser at B, and insert the wedges for tightening up the veneer.

If, then, everything is found to fit properly, the block may be released and *plenty* of glue laid on the joint preparatory to screwing and wedging up as before ; after which the screws should be inserted at B and the riser completed.

Care should be taken to see that no angle is formed at either springing, especially at the wedging, by driving the wedges too tightly. The straight part should be made perfectly tangential to the curve.

When the block is too large to be obtained conveniently in one piece, two or three may be used and well glued and screwed or nailed together, the grain of the wood in the different pieces being placed at right angles.

The plan of stairs as given in Fig. 1 will of course have to be varied to suit different circumstances, although one like this might do very well for some situations. Sometimes a whole half space, as from B to c, has to be occupied by winders, in other cases there are winders at bottom to start with, and in some instances winders at top to finish with. This last, however, should always be avoided in dog-legged stairs, if possible, as the narrow ends of the winders against the newel, when they are looked at from the floor-landing, give the stairs a very steep and dangerous appearance.

6. Beside the foregoing there are what is termed *open newel* stairs. Some very fine old specimens of the kind are to be met with in mansions and other places in different parts of the country, which, for substantial and noble appearance, combined with ease and comfort in ascending and descending, will far surpass a good many of the more modern productions. They were evidently designed by men having a thorough practical knowledge of the requirements as regards safety and comfort in such cases. They are used now, as they were in more ancient times, only in situations where there is plenty of room, and where expense is not, as in some instances at least, a consideration, as the stairs are designed and executed in a very elaborate and costly manner. Fig. 8 is a plan of such a staircase. This sketch, however, will give but a very poor idea of the appearance such a staircase presents when seen in elevation after being erected, especially if the materials and workmanship are of a first-rate kind. The handrails in such a case would usually have a swan's-neck easing at the upper end of the first and second run so as to

bring it up to the height of the level landing-
rails, and so make the same square in the newel do
for both.

But it is not often that an opportunity offers
of adopting such a plan for a staircase as Fig. 8
in private houses, the room required for such a
purpose being con-
sidered too great.
The landing, how-
ever, should always be
introduced wherever
it is possible to do so
in place of winders,
which in numberless
instances are not
only objectionable on
account of the little
going and perpen-
dicularity at their
narrow ends, but also
on account of the danger attending their use.
Thus, the stairs shown in plan at Fig. 1 would
be very much better with a half-space *landing*
instead of a quarter-space and a quarter of
winders as there shown. So also in the geometrical example as shown in plan at Fig. 9.
If it be possible when planning a stairs to
introduce a half-space *landing* in place of the

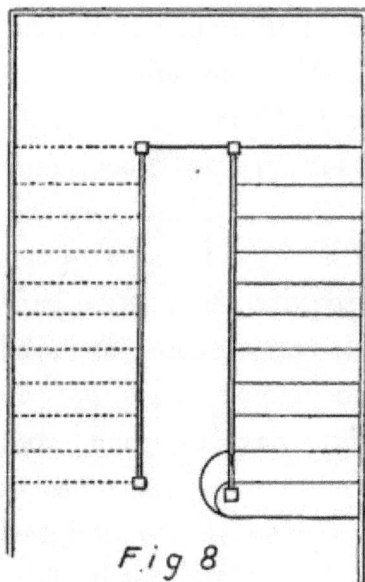

Fig 8

winders it should always be done, and the stairs will look better and be much safer and easier to use. These matters, as a rule, are decided by others than the workman who makes the stairs. Still, there are times when the stair-builder is called upon to exercise his judgment in reference to them, and it is for guidance in such cases that the author has ventured to offer the foregoing remarks.

7. The next example is a *geometrical* or *continued* staircase, but a good many of the remarks already made in reference to the dog-legged

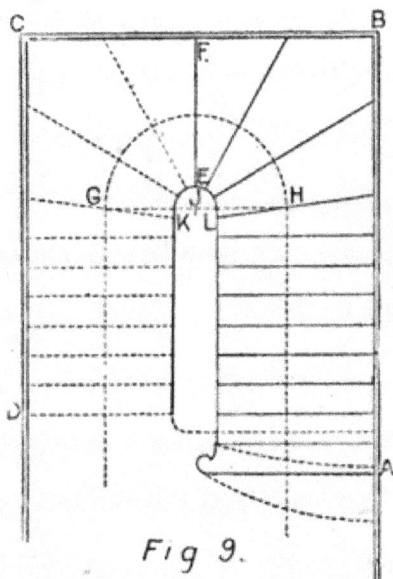

Fig 9.

staircases will be equally applicable to some particular parts of this also.

Let A B C D, Fig. 9, be the plan of the walls where a geometrical staircase is to be erected. The position of doorways either at the bottom or top, or wherever else they may occur if near the stairs, must be carefully noted, and the plan of the staircase laid down accordingly, as they should not be allowed too close to either the first or last step. The position of windows also has sometimes to be taken into account in laying down the plan. These, however, are considerations which will only affect particular cases as they occur in practice, and need not concern us here any further than as stated above.

A staircase such as we show in Fig. 9 is often required, and will generally come sufficiently clear of doors and windows to obviate any difficulty or necessitate any alteration when laying down the plan.

The same particulars stated to be necessary previous to planning a dog-legged staircase must also be determined in this case. Supposing these to be known, proceed to lay down the plan of the walls, as at A B C D, the width B C being equal to 7 feet, with plenty of *going* for the first and second run of flyers, and the height 11 feet 4½ inches. There being 21 risers in all, this will give a rise for each step of 6⅓ inches, for which a 10-inch *going* is a suitable proportion.*

Draw first the lines representing the thickness of the wall strings, and parallel to these the lines

---

* In deal stairs this *going* is very seldom called for, an 11 inch plank being the widest stuff available for the treads, and this will only allow of from a 9 inch to a 9¼ inch *going* at the very outside, which is amply sufficient for all ordinary purposes.

for the face of the outer strings, continuing the
same in a semi-circle for the plan of the well hole
and making the width at E F equal to the width
at the flyers, which is here supposed to be 3 feet,
thus making the *well hole* 1 foot wide. Through
the centre J draw the line G H parallel to B C, or
at right-angles to A B. G and H being the points
of division through which the lines of the diminished flyers are drawn, should be situated about
midway between the wall and outer string, or in
the supposed position a person would occupy in
using the stairs. Then, with a radius of 2 feet
describe the semi-circle as shown by the dotted
curve. Divide the semi-circle G H into the same
number of equal parts as there are to be winders
in the half space, not counting the steps G and H
as such, for although they have the appearance
of winders they are not really winders, their
proper name being *diminished flyers*. From the
points where the semi-circle E is cut by the line
G H set off along the face line of the outer strings
a distance equal to about 2 or $2\frac{1}{2}$ inches, as at K and
L. Then divide K E L into the same number of
equal parts as there are winders, and from these
points of division, and through the divisions in
the semi-circle G H draw the lines representing the
face of the risers. Also from K and L, through G
and H, draw K G and L H, producing them to the

wall string. Then from G and H set off the width
and required number of flyers along the dotted lines
as shown; and through the divisions draw the
riser lines at right angles to the face of the wall
string, finishing off at bottom with a curtail step,
and with a quarter circle at top as shown. The
size and shape of the curtail must be determined
by the method of drawing the scroll as given in
Part I. HANDRAILING, Plate 1 (5).

This method of setting out the winders will be
found to produce as good a nosing-line for the
narrow ends round the well-hole as can be desired.
Different methods of finding the nosing-line are
sometimes taught, by which the narrow ends of
the winders are made unequal in width and the
outline of the lower edge of the wreathed string
is wrought into the form of a large *ogee*, the concave curve meeting the convex about midway
between the springings.* This, however, will
not have so good an appearance as when the
lower edge is made as straight as circumstances
will permit, and by the method just pointed out
this object may be accomplished, in this and similar cases, by having a small easing at each end,
to bring it to the rake of the straight string.

Let any one note the appearance of a perfectly

* This is called by the French "making the steps dance."

straight edged piece of veneer, after being bent round a semi-cylinder, and glued up in the form of a wreathed string, in which the stretch out of the semi-circle is equal to the *going* of the tread, and he will perhaps have some idea of the author's meaning.

Modifications of the above must of course be made as required to suit particular cases as they occur in practice, so that no particular rule can be given which will apply to all.

A section of these stairs may be constructed in the same way as the section of the dog-legged example, and the length of strings and straight rail obtained for cutting off the stuff. The wall strings require to be set out in the same way as those of the previous example; the width of the housings for the winders being taken from the plan as they occur on the face of the string.

Fig. 10 shows the wall string for the winders.

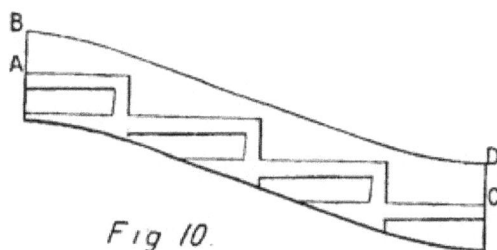

Fig 10.

The easing of the top edge may be made as considered most desirable. Fig. 11 shows the

lower end of the top wall string, the height A B
(Fig. 11), being equal to A B in Fig. 10.

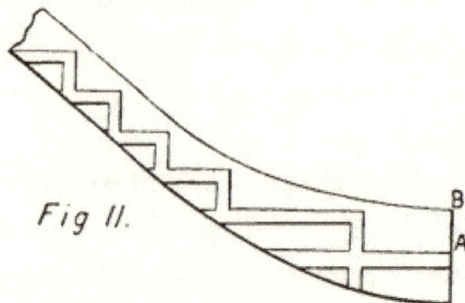

Fig II.

Fig. 12 shows the upper end of the first wall
string ; C D being equal to C D in Fig. 10.

These strings should be grooved and tongued
at the angles, and the bead or moulding on the
top edge mitred or scribed as found to be most
convenient.

We will now take the remaining operations

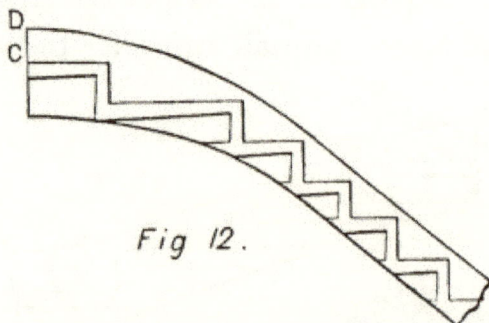

Fig 12.

relating to the outer string in the order in which
they occur as we ascend the stairs. This will
give us the curtail step first, which, as previously

stated, must be described by the method of drawing a scroll as given in " Handrailing." It will
be somewhat smaller than the scroll according as
the handrail overhangs the balusters little or
much ; the size at its smallest part, as A, Fig. 13,
which is called the *neck*, being about equal to the
size of the balusters. Fig. 13 shows the size of
the block, which should be of thoroughly dry
stuff and in about three thicknesses, well glued to-

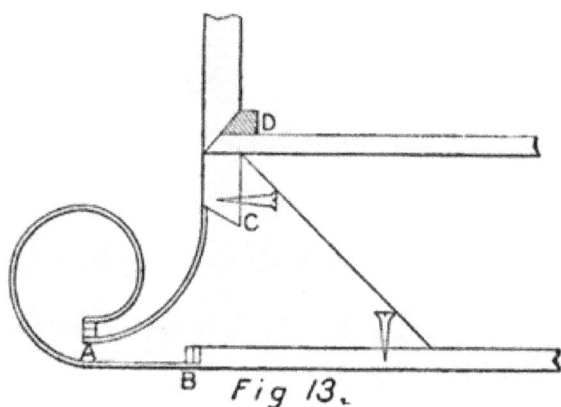

Fig 13.

gether ; the figure also shows the veneer attached,
and the position of the first and second risers.
The veneer should be put on according to the directions previously given for a round cornered step
(5), plenty of glue being used ; the wedges driven
in first at A and then at B, so that it may be
drawn up sufficiently tight, and made to lie perfectly close to the block, screws being used at the
thick part as shown.

The curtail may be connected to the string as at
c, and well secured to it from behind by screws.
At D the mitre for the second riser is shown,
which should be cut to an angle of 45°; and as
the string is usually one and a half times or twice
the thickness of the riser, a glued block should be
inserted as shown by the shaded section. Nails
may be driven through the string, especially if
painted, endways into the riser for bringing the
mitre together, but they should not be driven
through the face of the riser into the string
unless the risers are also painted.

The *scotia*, or hollow under nosing, of the curtail, is screwed or nailed on from above to the
top edge of the riser, as
s in Fig. 14 ; and is
made to project suffi-
ciently over the back to
allow of the tread being
screwed to it from un-

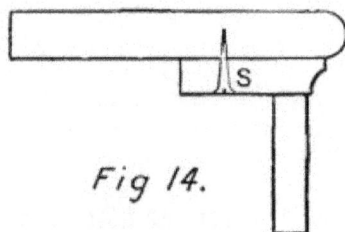

Fig 14.

derneath by screws as shown ; and the several
parts are well glued and screwed together after
they have been properly fitted.

The *scotias* for the flyers are let into a groove in
the treads, and the risers are then glued against
them and held in their proper position by glued
blocks at the back, as seen at G in Fig. 15. No
nails should be driven through the treads at the

front edge for this purpose. Sometimes the bottom edge of the riser
is tongued into the
back of the tread as
at A. Screws should
be used for securing
the back of the tread
to the lower edge of
the riser; and the

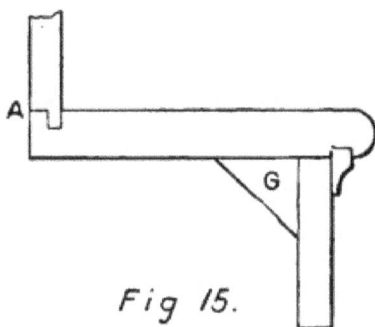

Fig 15.

joint, instead of being perfectly square, should be
shot a little under at the back.

The commode steps shown by the dotted curves
at the commencement of the plan in Fig. 9 may
be constructed as shown in Fig. 16, where A is a
supposed rib about 1½ inch thick, having one edge

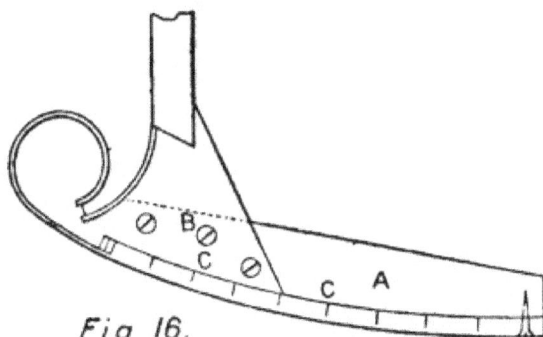

Fig 16.

cut to the curve of the back of the riser. This
rib is then let into the middle of the curtail block,
and when the curves of each are properly adjusted, the rib and block are glued and screwed

together as at B. The thick part of the riser is next *saw-kerfed* on the back, as at c c, until it can be bent to the form of the rib, and held there by a handscrew, or by an ordinary screw driven through close to the end.

Then a plentiful supply of glued blocks are applied on both sides of the rib, and to the back of the riser ; and the angles formed on the face opposite the *saw-kerfs* are eased off with a finely set smoothing plane. The remaining operations are as described for Figs. 13 and 14.

CHAPTER II.

THE WREATHED STRING.

8. THE next subject for consideration is the
" wreathed string " for the narrow ends of the
winders at K E L, Fig. 9. But in the first place
we must consider the *cylinder*, or more properly
semi-cylinder, on which the wreathed string is
glued-up to its proper form. The semi-cylinder
is prepared as shown on Fig. 17,
the width, K L, being made equal
to the width or diameter of the
well hole, or the distance in the
clear between the faces of the
strings, K L (Fig. 9). The semi-
circle, unless it is larger than
usual, is formed with two pieces
of the required thickness, and
the two straight sides of 1-inch
board about 18 inches wide ; the whole to be nailed
to two legs cut to the required shape at top. The
length should be sufficient to take the whole length
of the string that is required to be glued-up.

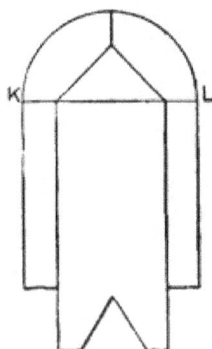

There are several ways in which the wreathed

string for the well hole is prepared, but we will explain first the one most generally adopted, and considered by competent authorities by far the best. It is capable of the highest finish, can be applied to all cases, and is as strong as it is possible to make a string for such a purpose in wood. A thin veneer cut out of yellow pine is the material most usually employed when the work has to be painted, and when not to be painted it must of course be of the same kind of stuff as the string. This is placed on the semi-cylinder, as we shall presently explain, and backed up with pieces of pine, or white deal, $2\frac{1}{2}$ inches or 3 inches wide, hollowed out to fit the curve.

To obtain the correct size and shape of the veneer most readily, it will be found best to develop the semi-circular part of the well hole and the steps and risers connected therewith on a drawing-board first. The necessary lines can then be transferred from the board to the veneer rapidly and correctly, and its size and shape determined at once.

Proceed first to take a thin slip of wood, or other flexible material, and wrap it closely round the semi-circular part of the cylinder, marking on it the distance from springing to springing while so applied, and draw two parallel lines at a perpendicular distance apart equal to the distance from springing to springing of the cylinder just taken by the slip of wood, as A B and C D on Fig. 18.

The face of all the risers will be parallel, and the treads at right angles to these two lines. If, therefore, the positions of the treads and risers are

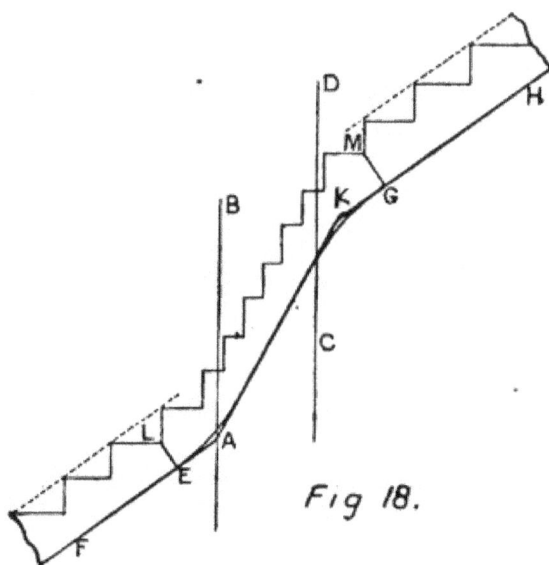

Fig 18.

carefully noted as they occur on the face of the string in the plan Fig. 9, their correct positions in the development on Fig. 18 will be easily determined. Transfer these as required to Fig. 18 and also one or two flyers above and below as shown. Draw the lines E F and G H parallel to the nosing-line of the flyers and at a distance from it, equal to the width of the string, producing the lower one to cut the springing A B in A. From A draw A K parallel to the nosing-line of the winders and cutting G H produced in K. Ease off the angles as shown by a small easing at both ends, and through the inner angle of the tread and riser next but one to the springing, as at L and M, draw the lines E.L and G M square to E F and G H. This gives us the joint lines between the straight and wreathed strings, and determines the size and shape of the veneer as from E to G. This will complete the development which must now be transferred to the face of the veneer; but the indentations for the steps, and the easing on the lower edge should not be cut until after the wreath has been properly formed by glueing it up with the backing on the cylinder, and jointing it to the strings.

The straight parts of the veneer beyond the springing on either side, as at E and G, are called the *tangents*, and each of these will require to be backed up with about two pieces in width, each two when jointed together being wide enough to cover the *tangent;* these are called the *tangent-backing-pieces*, and the material of which they are made should be thoroughly dry and sound, of the same thickness as the string, the joint and grain running parallel to the springing lines, and one edge of each tangent-backing-piece must be placed exactly to the springing line on the veneer.

Before attempting to place the veneer on the cylinder, the tangent-backing-pieces just specified should be well glued on to the back of the veneer and left to dry. These backing pieces should be long enough to reach beyond the top and bottom edges of the veneer. When the tangents are sufficiently dry, about two holes should be bored through, and one of the tangents should then be screwed to one side of the cylinder, the springing lines of the two being placed exactly opposite each other. Then the veneer should be gently bent round the cylinder, and the other tangent screwed to the opposite side in a similar way. The veneer is then ready for the backing, which is made in pieces from 2 inches to 3 inches wide, and hollowed out on one side to correspond with the semi-circular back of the cylinder as shown in Fig. 19, with the joint lines pointing in the direction of the centre of the curve. In glueing

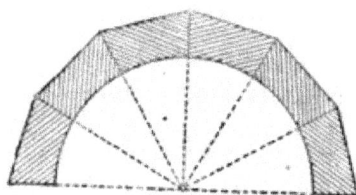

Fig 19.

these pieces on to the veneer, they should be started from one of the tangents on either side, and each piece should be long enough to allow of a screw or nail being driven through at each end to hold it securely in its place while the next one is being put on.

Plenty of glue must be used and each of the joints well rubbed so as to make sure of filling up all the interstices and making the whole into one perfectly solid piece. The wreath may remain on the cylinder long enough for the glue to get fairly set, but not until it is dry and hard. It should then be taken off by withdrawing the screws and nails, and allowed to get thoroughly dry before the workman attempts to connect it to the strings.

The joints at E and G (Fig. 18) are easily made by squaring them from the lower edge, but in addition to this a large bevel should be set with its stock to the line E F, or edge of the straight string, and its blade to the springing line A B. Then in trying the joint together it must be adjusted so that the springing line on the wreath coincides with the blade of the bevel, while its stock is held to the edge of the string. The upper joint is made similarly, but the stock of the bevel is applied to the top instead of the bottom edge of the string. For the purpose of securing and strengthening the joints when made, the simplest and most effectual method is to have a piece of $1\frac{1}{2}$-inch stuff about 4 inches wide and from 2 to 3 feet long as may be required. This is well glued and screwed to the back of the straight end

of the wreath, as at A, Fig. 20, which is a view
of the strings and piece edgeways; a hardwood
key piece about 3 inches
by ⅓ inch thick is then
glued and screwed on to
the straight string, and
the stiffening piece notched
out to fit over it, a pair of
wedges being inserted as
shown for tightening up
the joint. When this end
has been thus properly fitted
it is, with the wreath attached, glued and screwed
to the string, as shown at
Fig. 20. This can be ap-
plied to all joints of a
similar description, and if
well done will make the joint a very secure
one.

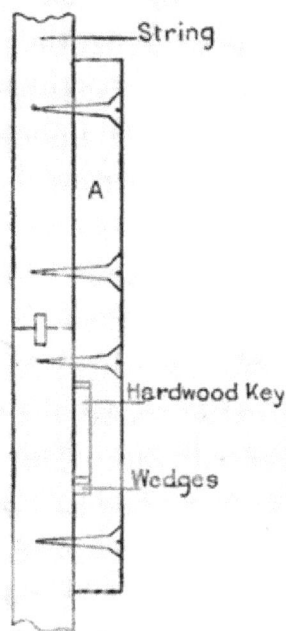

The indentations for the steps are best left, and
cut one at a time as required when the steps are
being fitted into their places, especially if this is
done at the building, as the sharp arrises for the
mitres may thereby be preserved intact and the
work made to look well when done.

9. The form of the wreathed-string for the
quadrant at top of the staircase may be found

similarly to the one just described by developing the quarter-circle on a drawing board, and one or two flyers at the lower end so as to obtain the proper rake for the lower edge of the string. This should be brought to the level of the under edge of the fascia lining for the trimmer by a small casing. One flyer should be included in the wreath at its lower end as in the semi-circular string, and the tangent and joint glued up in exactly the same way. The tangent at the top end may be made any length from 6 inches to 1 foot as desired, and rebated out on the front for the thickness of the fascia lining. The cylinder used for the winders is made to do by fixing a straight piece over one half of it flush with the top of the semi-circle temporarily, and at right angles to the side. The wreath, prepared as before directed, is then placed on it, springing to springing, and the backing pieces glued round the circular quadrant as previously. The position of the last riser should be found as directed in " Handrailing " (7).

10. It is a practice with some stairbuilders to glue a piece of coarse canvas on the back of such strings as the foregoing, although it may not be really necessary in all cases; but it undoubtedly tends to considerably strengthen them when used, and as it involves very little extra trouble and expense it should be done whenever extra strength

is required, or when the wreath is a very long one.

11. When the string has a sinking the wreath should be made wide enough to include it, and the cutting line must then be put on at the depth of the sinking, as at A, Fig. 21. A plough-grove may then be made in the under edge of the upper part of the string, and a tongue worked on the top edge of the piece cut off to form the rebate, and with a little careful manipulation this lower part may be set back into the groove the required distance, and the sinking thus formed without any difficulty.

The return bead on the bottom edge should be made of cane, bent gradually round and closely bradded.

Fig 21

12. The method described above for finding the lower edge of the wreathed string may require some trifling modifications for different plans. It need not always at every point follow the line of the winder nosings in strict parallelism, neither should it be equal in width to the width of the straight string when measured perpendicularly to the falling lines. The depth should rather be measured and made equal, each to each, as nearly as practicable in a perpendicular direction at the

face of the risers. If, however, when so measured the wreathed string is made somewhat deeper, by means similar to those pointed out, than the straight string, it will be found to have a much better appearance than when measured and made equal in width the opposite way.

Every workman having the least acquaintance with the subject will know perfectly well that these kinds of strings always present a very different appearance when wrought into their proper form than when they are simply laid out on a flat surface, as in the development shown on Fig. 18. It, therefore, follows that, provided a sufficient depth of string is secured to cover a sufficiently strong carriage, it is to their appearance when finished that the greatest attention must be paid, and not so much as to how they look when laid out as in the development on a flat surface.

13. The "carriages" are placed immediately under and parallel to the risers. They are let into the wall and well wedged at the back end, and nailed to the wreathed string at the other end. *Cross-bearers* as a rule are not necessary, they only add to the weight without being of any service. The carriage pieces already referred to being under the back edge of the winder, always cross the grain sufficiently without any additional cross-bearers from carriage to riser.

14. Another method of preparing a wreathed string, is to hollow out a solid piece of stuff on one side to the curvature of the well hole, and in good-sized wells to glue several pieces together so as to obtain the desired curve ; the grain of the wood running in a vertical direction. The piece as thus prepared is then jointed to the straight strings, the indentations for the steps are set out and cut, and the easing on the lower edge made to follow the angles of the steps as nearly as may be considered desirable.

With a very small well hole, and where no winders are required the above method might do very well, but for any other situation it is not to be recommended at all, although it is sometimes made to do, but only in an inferior class of work.

15. Another method is sometimes practised when the curved surface is of great length and large sweep, as in the wall-strings of a staircase having a circular plan. In this method two ribs are prepared so as to fit as much of the circular wall as may be required when set up to the rake of the stairs. They are then set to the proper angle at their ends and the curved edges covered with boarding, thus forming a cylindrical surface of elliptical curvature.

H

Fig. 22, shows a semi-circular * plan of the walls for a staircase of this description, and Fig. 23 represents the ribs elevated to their supposed pitch, and showing their straight edges with the boarding going round the curved edges behind. The ends, c and d, are of course vertical when the cylinder is set up to its correct pitch. Fig. 24 shows one of the ribs flatways, with the boarding on the curved edges. The string is made up of several thicknesses of veneer sufficiently thin to bend easily; two or three pieces being glued up at a time, with bars of wood placed across them sufficiently close, and screwed at each end into the boarding of the cylinder; so as to force out the glue and press the joints up closely. The veneers should not be too thick, or there will be a

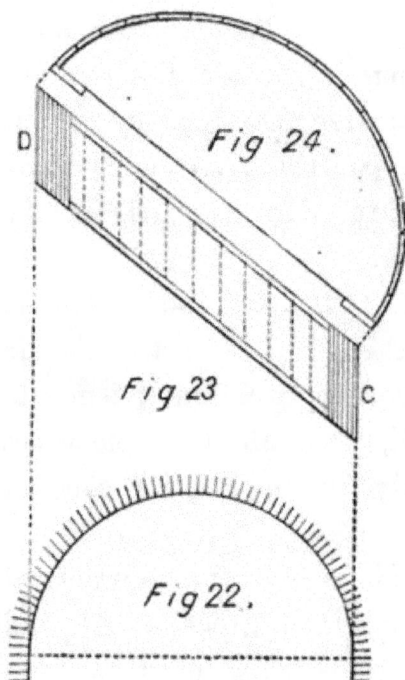

---

* The same method is adopted when the stairs occupy only the segment of a circle.

strong tendency to spring back to straightness when taken off the cylinder.

16. A very strong and substantial form of string for similar uses to the foregoing, and for an outer string when such is required of similarly large sweep, may be made up on the foregoing cylinder. The method is to use circular bars of wood about 2 inches wide and of any convenient length, cut out of the solid to a concave curve on one edge that shall fit the back of the cylinder and to the required parallel thickness, the pieces being straight the opposite way.

The pieces are fitted on the cylinder one at a time, the concave edge being worked to the slight twist that will be required, and the straight sides placed parallel to the straight sides of the cylinder. They are then glued and screwed together on the cylinder in sufficient numbers to make up the width of string required, and breaking the heading joints wherever they occur. When the glue is thoroughly dry and hard, both sides, if for an outer string, and one side only, if for a wall string, may be straightened up in a vertical direction, preparatory to covering one or both sides as required with a veneer of the requisite thickness.

CHAPTER III.

Bracketted Stairs.

17. In the better kind of staircases the outer string is frequently ornamented with brackets of various fancy patterns, which are mitred to the riser and glued and bradded to the face of the

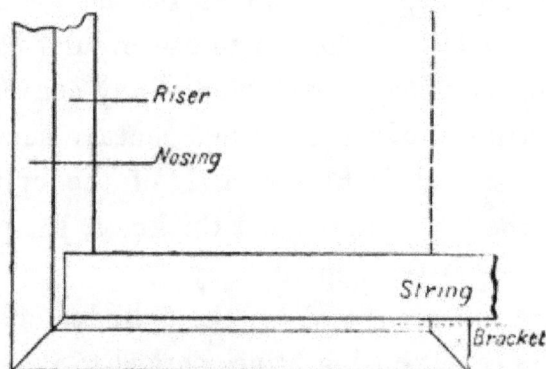

Fig 25.

string, the tread projecting far enough over to include the thickness of the bracket.

Fig. 25 is a part plan showing how the riser is shouldered to the back of the string and the bracket mitred to it; the nosing of the tread and the return nosing being also shown.

Fig. 26 is a part elevation showing the face of

the string with the bracket planted on and mitred to the riser, and the return nosing on the end of the tread.

. Sometimes these brackets have panels formed in them by shallow sinkings, or they may be ornamented in a variety of other ways.

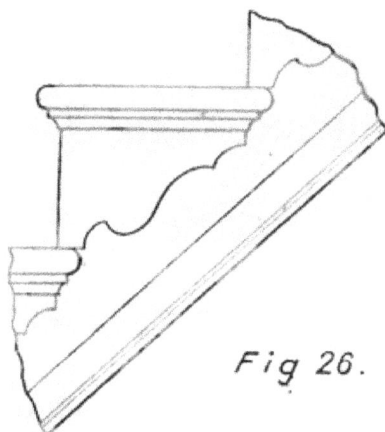

Fig 26.

18. ENLARGING AND DIMINISHING BRACKETS.

The pattern bracket usually given is of a size suitable for the end of a flyer; but when winders occur in the stairs this pattern must be diminished or enlarged to suit the varying widths of the same.

To do this make the width of the given bracket the base of a right-angled triangle, as at A B, Fig. 27; then setting off any convenient distance, as A C, for a perpendicular, draw the hypothenuse, B C, of the triangle, as shown. Draw D E the width of the shorter bracket, which is required, parallel to A B.

In the moulded edge of the original bracket take any number of points, and from them draw ordinates to cut A B; then from these points in A B, draw lines converging to the point C. Then from the points where D E is cut by these converging lines erect perpendicular ordinates as shown, and make them respectively equal to the corresponding ordinates in the original bracket. Then, by tracing the contour through the points thus obtained the bracket will be produced.

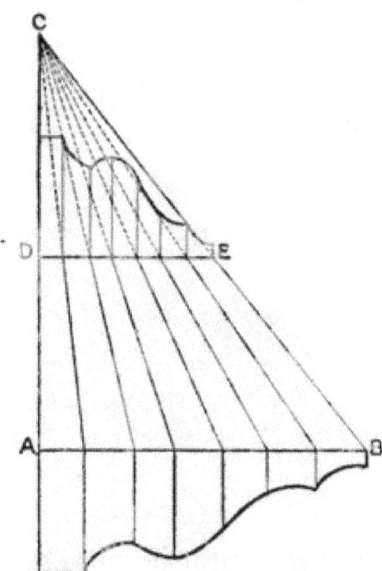

Fig 27.

The above method is for diminishing. When it is required to enlarge a bracket, it is only necessary to reverse the process by making the narrower bracket, as D E, the base of the triangle, and producing from it the perpendicular and hypothenuse far enough to obtain the width of the bracket to be enlarged. The procedure in this case will be sufficiently obvious without a detailed description.

PRINTED BY J. S. VIRTUE AND CO., LIMITED, CITY ROAD, LONDON.

www.ingramcontent.com/pod-product-compliance
Lightning Source LLC
Chambersburg PA
CBHW021814190326
41518CB00007B/583